FLAVOR RESEARCH

Principles and Techniques

FOOD SCIENCE

A Series of Monographs

Series Editor

OWEN R. FENNEMA
DEPARTMENT OF FOOD SCIENCES & INDUSTRIES
UNIVERSITY OF WISCONSIN
COLLEGE OF AGRICULTURE
MADISON, WISCONSIN

FLAVOR RESEARCH: Principles and Techniques
by R. Teranishi, I. Hornstein, P. Issenberg, and E. L. Wick

OTHER VOLUMES IN PREPARATION

FLAVOR RESEARCH

Principles and Techniques

ROY TERANISHI

Fruit Laboratory
U. S. Department of Agriculture
Albany, California

IRWIN HORNSTEIN

U. S. Department of Agriculture
Beltsville, Maryland

PHILLIP ISSENBERG

Department of Nutrition and Food Science
Massachusetts Institute of Technology
Cambridge, Massachusetts

EMILY L. WICK

Department of Nutrition and Food Science
Massachusetts Institute of Technology
Cambridge, Massachusetts

MARCEL DEKKER, INC., New York   1971

INTRODUCTION TO THE SERIES

All fields of science must be nourished by a systematic input of new knowledge and refined principles if proper growth is to occur. Books are a major means to this end. Food Science because it is young and afflicted by growing pains remains undernourished with respect to high-quality books. This is especially true of books for advanced undergraduate students, graduate students, teachers, and persons involved in research and development. FOOD SCIENCE MONOGRAPHS are directed primarily to these needs.

My personal philosophy, as consulting editor, should be mentioned because it has a bearing on selection of authors and the general character of the series. Because Food Science is an interdisciplinary science, it is important that authors utilize a thoroughly-integrated mixture of conventional food science data and data from allied sciences. This enables an adequate fundamental background to be laid for presentations of technological principles, and if done properly, is the key to imbuing readers with an in-depth understanding of the subject. This approach, although difficult, is especially well-suited to the needs of the prospective audience. An important question concerns the degree to which this philosophy can be implemented. I am confident of success--success as measured by professional benefits to the audience described, and by advancement of Food Science in general. My confidence is fully justified, since it is based on the outstanding abilities of those who have agreed to participate in this series.

Owen Fennema
Consulting Editor

Madison, Wisconsin
1971

iii

PREFACE

Our lack of knowledge concerning the mechanisms by which we perceive tastes and odors contrasts vividly with our understanding of the processes by which we perceive sights and sounds. The technology developed to exploit the understanding of these phenomena reflects this difference. We can record, store, retrieve, amplify, transmit, duplicate, and describe objectively the sights we see and the sound we hear. Incredibly, none of these operations can be duplicated for a single taste or a single odor!

The chemists' approach to the study of food flavors is empirical. The painstaking isolation and identification of key compounds must be carried out before a naturally occurring flavor can be duplicated. Side-benefits of these studies are the accumulation of data that may lead to an understanding of the relationships between chemical structure and taste and odor. This empirical approach may also help uncover the clues that can lead to an understanding of how the chemical senses of taste and olfaction function.

For man flavor and nutrition are opposite sides of the same coin. Our senses of taste and smell permit us to enjoy food and drink. Without this enjoyment man will reject even the most nutritious and wholesome food.

Experience, and perhaps to a degree instinct, lead us to associate "pleasurable" odors and tastes with foods that are "good." Once formed these associations are difficult to break. As the population increases, our caloric and protein requirements inevitably will be supplied by unconventional foods. Even in the most developed countries, the luxury of converting cereals to meat may become untenable. Protein foods derived from petroleum, oilseeds, legumes, and the like may be man's dietary lot. The persistence of food preferences for the flavors of traditional foods may prove a major roadblock to the utilization of these new foods. To incorporate into fabricated foods, flavors acceptable to populations differing in their food likes will add a new urgency to the flavor chemists efforts.

Flavor studies emphasize the volatile compounds that are responsible for food aromas. Trace amounts of volatiles in some instances at the fantastically low levels of parts per trillion, can have profound effects on odor sensations. To study odors and olfaction ultramicro methods must be utilized. Until recently, neither the instrumentation nor the techniques were available. Now complex, delicate, and expensive instruments are being used to sift nonsense from facts, and to bring orderliness and chemical sense to the study of flavor.

Of equal importance is accumulation of detailed knowledge of food composition; in particular, the influences of processing, additives, packaging, and storage. Food is a major component of man's chemical environment. The nature and distribution of trace organic food constituents are as important to his well being as is the quality of air and water consumed.

Flavor research is an active and rapidly changing field. No doubt by the time this book reaches the reader many advances will be made. It is the hope of the authors that this discussion of advantages and pitfalls in using available techniques, methodology and instrumentation, will provide short-cuts to meaningful results in this fascinating field of food and flavor research.

The authors thank Mr. Robert David Wong, Scientific Information Specialist, and his staff in the Illustration Studio of Western Regional, for the artwork and preparation of the figures for camera-ready copy.

CONTENTS

Chapter 1

CHEMICAL ASPECTS OF FLAVOR

I. INTRODUCTION

All our senses contribute to the flavor of a food. Color, texture, taste, odor, and even sound (picture eating silent celery or crackling spaghetti!) are all part of the flavor profile. A definition of flavor

1

recently proposed by Hall (1) and representing a consensus of expert opinion limits flavor to experiences based on oral contact: "Flavor is the sensation produced by a material taken in the mouth, perceived principally by the senses of taste and smell, and also by the general pain, tactile and temperature receptors in the mouth. Flavor also denotes the sum of the characteristics of the material which produce that sensation."

This definition stresses the dominant roles of taste and smell but does not assess their relative contribution. One hundred and forty-five years ago Brillat-Savarin (2) wrote: "He who eats is conscious of the smell of what he is eating either at once or upon reflection, and towards unknown foodstuffs the nose acts as an advanced sentry crying 'who goes there?' When smell is intercepted taste is paralyzed ..."

It is the importance of odor to flavor so well phrased by this 18th century gourmet, plus the development of techniques particularly suited to separate and identify trace amounts of volatile compounds, that have been instrumental in promoting the investigation of flavor during the past 10-15 years.

The flavor chemist has labored mightily to identify volatile compounds. Perhaps too often, the pursuit of identity has been more enthusiastic than critical. Isolation procedures have introduced artifacts that vitiated results, separation techniques have yielded mixtures where purity was sought, and dubious evidence has produced identifications that are only slightly probable.

This book is primarily devoted to a critical examination of the rationale, techniques, and methodology that have been developed to study volatile flavor compounds. The introductory chapter summarizes briefly some relevant aspects of current knowledge (or lack thereof) concerning taste and smell, and details the uniqueness of the problems facing the flavor chemist. Some references peripheral to the main purpose of the book, but germane to the field of flavor research, are given at the end of this chapter.

II. TASTE

Estimates of the number of basic taste qualities have varied from a minimum of two to a maximum of eleven. Linnaeus (3) in 1754 postulated eleven basic tastes: sweet, sour, sharp, salty, bitter, fatty, insipid, astringent, viscous, aqueous, and nauseous. A century later the number had dropped to two: sour and salty. Current usage, admittedly inadequate, is that sweet, sour, salty, and bitter are the four basic tastes.

A. Receptors

Pfaffman (4) investigated the degree of specificity of taste receptors. Working with rats and hamsters he recorded the electrical neural response of single nerve fibers and found that a single fiber could respond to more than one basic taste stimulus. Some nerve fibers exhibited broad sensitivity to sugar, salt, quinine, and hydrochloric acid; other fibers responded to salt, sugar, and acid; still others to salt and sugar and the like. There was no indication that specific receptors existed for any of the basic taste qualities. Rather the pattern of nerve activity appeared to depend on the nature of the chemical stimulus. The response magnitude of single nerve fibers to the same stimulus also varied from fiber to fiber. These results were not wholly conclusive since single nerve fibers may respond to more than one taste cell. Kimura and Beidler (5) recorded the electrical responses from single taste cells and obtained results similar to those from nerve fiber studies. The conclusion seems inescapable that our perception of the basic taste qualities results from a pattern of nerve activity coming from many taste cells and that specific receptor cells for sweet, sour, bitter, and salt do not exist. It may be envisioned that a single taste cell possesses multiple receptor sites each of which may have specificity. Variations in the number and kind of

receptor sites from taste cell to taste cell could produce qualitatively
and quantitatively different responses to the same stimulus.

Knowledge concerning the nature of the taste receptors is meager.
Dastoli and Price (6), by means of ammonium sulfate fractionation,
isolated a crude protein fraction from bovine tongue epithelium having
properties expected of a sweet taste receptor molecule. Their assay
method measured a shift in refractive index of their protein fraction
when sugars were added. This shift in refractive index was assumed
to result from a protein-sugar complex formation. The equilibrium
constant for the binding of sugars by the "sweet-sensitive protein"
corresponded to the order of sweetness observed for these sugars in
humans. An implied but unproved assumption underlying these studies
is that man and beast respond alike to different sweet-tasing substances.
Hiji et al. (7) have also reported sugar-complexing activity in protein
fractions isolated from rat tongues. Cats are relatively insensitive
to sugars. Interestingly enough, extracts of cat tongue epithelium
contain very little sugar-complexing protein (7).

Dastoli et al. (8) also reported the isolation of a protein fraction
that complexed bitter-tasting compounds. That this protein is actually
the bitter receptor has been questioned (9). The evidence that the
sweet and bitter taste receptors are proteins must, as of now, be
considered inconclusive.

B.   Mechanism of Taste Initiation

Quantitative measurements of electrical neural response as a
function of stimulus concentration show that the response increases,
but at a decreasing rate, as stimulus concentration is increased. A
point is finally reached where further increases in concentration
produce no increase in response. Beidler (10) analyzed such data and
derived an equation relating magnitude of response and stimulus
concentration:

$$\frac{C}{R} = \frac{C}{R_s} + \frac{1}{KR_s}$$

where C is stimulus concentration, R is response magnitude, R_s is the maximum response, and K is the equilibrium constant for the stimulus-receptor reaction. This equation yields a sigmoid curve when plotted semilogarithmically (R vs. log C). Beidler's calculation of equilibrium constants for many substances yielded K values in the range 5-15.

As a result of these and additional taste receptor studies (11) Beidler concluded:

(a) The initial event is the adsorption of the stimulus on the surface of the taste receptor to form a weak stimulus-receptor complex. This disruption of the surface leads to the initiation of the nerve impulse.

(b) The number of receptor sites is finite.

(c) The taste response is a function of the fraction of sites occupied by the stimulus compound.

The concept that adsorption is the key initial step is in line both with the small values of K, which indicate that only weak interactions are involved, and with other data indicating that taste responses are relatively insensitive to pH changes and temperature changes.

C. Structure and Taste

The salt taste is typical of inorganic salts such as sodium chloride, ammonium chloride, and potassium chloride. The higher molecular weight salts such as cesium chloride or potassium iodide are bitter. To a first approximation one can state: (a) An ionized salt of low molecular weight usually gives a salty taste; (b) with increasing molecular weight there is a shift in taste from salt to bitter; (c) both anions and cations contribute to the salt taste. Thus, sodium chloride and sodium sulfate differ in "saltness," and sodium chloride and potassium chloride are not identical in taste.

The sour taste is a function of pH. Mineral acids at the same pH are equally "sour"; however, the taste of organic acids is affected by the organic moiety, and the simple relationship between pH and "sour-ness" is not clear-cut.

The understanding of relationships between structure and the sweet taste is limited. Shallenberger (12) and Shallenberger and Acree (13) have studied the influences of hydrogen bonding, stereochemical configuration, and conformation on the sweetness of hexoses and have enunciated rules for predicting the structures of compounds that might taste sweet.

Other than stating that alkaloids and glycosides are bitter, no general relationships have been established that correlate structure and the bitter taste.

In general, sweet and bitter-tasting compounds have been found in almost every class of organic compound. Further, relatively minor changes in structure may convert a sweet-tasting compound into a tasteless or bitter compound. The examples in Fig. 1-1 are illustra-tive: 2-amino-4-nitro-propoxybenzene (1) is 4000 times sweeter than sucrose, the 2-nitro-4-amino-propoxybenzene (2) is tasteless, and the 2,4-dinitro-propoxybenzene (3) is bitter. Dulcin (p-ethoxyphenyl-urea) (4) is extremely sweet, while the thiourea analog (5) is bitter and the o-ethoxyphenylurea (6) is tasteless. Saccharin is 300 times sweeter than sugar, but the N-methyl derivative is tasteless, etc.

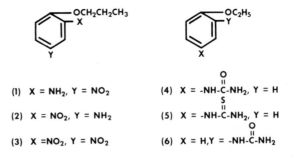

(1) X = NH$_2$, Y = NO$_2$ (4) X = -NH-C(=O)-NH$_2$, Y = H

(2) X = NO$_2$, Y = NH$_2$ (5) X = -NH-C(=S)-NH$_2$, Y = H

(3) X =NO$_2$, Y = NO$_2$ (6) X = H,Y = -NH-C(=O)-NH$_2$

FIG. 1-1 Some sweet, bitter, and tasteless compounds.

Stereoisomers may have widely different tastes. Solms et al. (14) reported L-tryptophan to be half as bitter as caffeine and D-tryptophan to be 35 times sweeter than sucrose. L- and D-phenylalanine and L- and D-tyrosine were reported to have similar but less intense taste qualities. The anomers of D-mannose are reported to differ in taste; ∝-D-mannose is sweet while β-D-mannose is reportedly bitter. Kulka (15) has reviewed the literature on structure-taste and cites many examples for which limited generalizations have been formulated. For example, piperine is largely responsible for the "bite" of black pepper. Staudinger and Schneider (16) found that to give this desired bite, the molecule must contain a phenyl group, a side chain of at least four carbon atoms, and the acid amide group derived from piperidine. The methylenedioxy group present in piperine as well as the two double bonds in the side chain are unessential.

Piperine

D. Enhancers

The addition of the sodium salt of L-glutamic acid to foods, particularly meaty and brothy foods, produces a flavor change that exceeds the flavor contribution of the pure compound. This phenomenon is exhibited by other amino acids such as tricholomic acid and ibotenic acid, isolated from Japanese fungi by Takemoto and Nakajima (17). The 5'-nucleotides are another group of naturally occurring compounds that exhibit this marked flavor-enhancing property. The sodium salts of 5'-inosine monophosphate and of 5'-guanosine monophosphate are the most effective. These are particularly effective in improving the flavor of meaty and soup-like products. Their structures are shown in Fig. 1-2. Combinations of these nucleotides with sodium glutamate

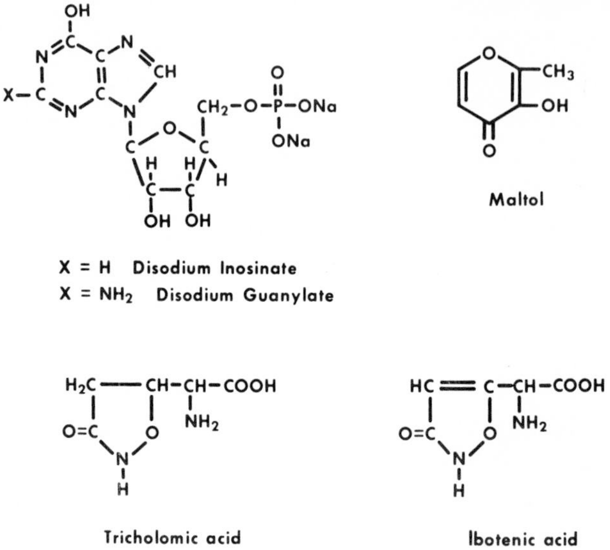

X = H   Disodium Inosinate
X = NH₂   Disodium Guanylate

Maltol

Tricholomic acid

Ibotenic acid

FIG. 1-2 Some flavor enhancers.

are reportedly even more effective than the single compounds alone (18).

Recognition that these compounds possess flavor-enhancing properties arose from flavor studies of typical Japanese foods. In 1909 monosodium L-glutamate was isolated from sea tangle, a kind of seaweed, and in 1913 the histidine salt of inosinic acid was isolated from dried bonito (18). Monosodium glutamate has been utilized commercially almost from the day of its isolation from seaweed. Because of production difficulties the 5'-nucleotides were not available in quantity until the early 1960's. Tricholomic acid and ibotenic acids are apparently not being used as enhancers.

Kuninaka (18) studied the flavor-enhancing activity of these potentiators alone and in combination. His results indicated that (a) 5'-inosinate and 5'-guanylate have the same kind of flavor-enhancing action, and a simple additive relationship exists when they are used together; (b) monosodium L-glutamate, tricholomic acid, and ibotenic acid exhibit similar types of flavor-enhancing action, different from that of the inosinates, and when used together these three compounds also exhibit an additive relationship; (c) when a compound from group (a) (an inosinate) and a compound from group (b) (an amino acid) are used together a greater than expected flavor enhancement is observed.

The neurophysiological effects of glutamate and the inosinates have also been studied (19). The electrical responses from single taste nerve fibers have been recorded and response patterns similar to those obtained from other taste-evoking compounds were observed. The synergistic action between L-glutamate and 5'-inosinate was also confirmed when mixtures of L-glutamate and 5'-inosinate, each at less than their respective threshold levels, yielded significant taste responses (19).

Maltol is used to enhance sweetness in such products as soft drinks and baked goods (15).

The mode of action of these synergists is unknown and the search for new enhancers must be on an empirical basis. A systematic screening campaign would undoubtedly yield many more compounds of possible potentiating value.

E. Inhibitors

Perhaps the most fascinating tools that may lead to both an understanding of the mechanism by which taste sensations are triggered and to the decoding of structure-taste correlations are the compounds that inhibit or change our perception of taste qualities. Two such materials have been studied in detail; one suppresses the ability to sense sweetness, the other changes the perception of sour to sweet.

If one or two leaves of a tropical plant, Gymnema sylvestre, R. BR., are chewed, the ability to taste sweet substances vanishes. Sugar placed in the mouth resembles sand. Taste sensitivity to saccharin,

cyclamate, glycerol, etc., also disappears. These effects may last
for several hours. To a degree, sensitivity to bitterness is also
diminished.

Hooper (20), one of the first to study the nature of this substance,
identified the active principle as being acidic and coined the name
gymnemic acid. Warren and Pfaffman (21) isolated a crystalline
gymnemic acid and suggested that it was a glycoside. Stöcklin et al.
(22) separated gymnemic acid by thin layer chromatography into four
components which they designated as gymnemic acids A_1 (the major
component), A_2, A_3, and A_4. The four compounds were D-glucuronides
of different acylated gymnemagenins. Based on IR, UV, NMR, and
mass spectral data, the unacylated gymnemagenin was assigned the
structure of a hexahydroxy pentacyclic triterpene.

Gymnemagenin

It is not yet known if the sugar part contains one or more molecules
of glucuronic acid and if the glucuronic acid is partly esterified (23).
The acids identified after hydrolysis of the mixed aglycones were
formic, acetic, butyric, isovaleric, and tiglic. Kurihara (24) reported
that the acids produced on hydrolysis of gymnemic acid A_1 were 1
mole of acetic acid, 1 mole of tiglic acid, and 2 moles of isovaleric
acid. He also reported that gymnemic acid A_1 could be converted to
A_2 by removal of 1 mole of tiglic acid and 2 moles of isovaleric acid.
Stöcklin et al. (22) had identified five acids but these were from the
mixed gymnemic acids.

Kurihara (24) also studied the effect of gymnemic acid A_1, the major component and most effective inhibitor, on a variety of normally sweet-tasting compounds including sucrose, cyclamate, D-tryptophan, D-leucine, beryllium chloride, lead acetate, and chloroform. All of these compounds except chloroform lost their sweet taste, an indication that at least two different pathways for perceiving sweetness may exist.

After one chews the pulp of miracle fruit, a small red berry of the Nigerian plant <u>Synsepalum delcificum</u>, a normally sour-tasting substance will instead taste sweet. This action was first described by Daniell (25) and years later investigated by Inglett et al. (26). Their efforts to isolate the active material were unsuccessful, primarily because of their inability to solubilize the active principle. Kurihara and Beidler (27) found that the active principle could be extracted from a homogenized slurry of these berries at pH 10.5. The active material was shown to be a glycoprotein whose molecular weight is approximately 44,000. The sugars were identified as L-arabinose and D-xylose. The amino acid composition is given in Table 1-1.

TABLE 1-1

Amino Acid Composition of Taste-Modifying Protein[a]

Amino acid residue per 100 total residues		Amino acid residue per 100 total residues	
Lysine	7.9	Alanine	6.3
Histidine	1.8	Cysteine	2.3
Ammonia	17.4	Valine	8.0
Arginine	4.7	Methionine	1.0
Aspartic acid	11.3	Isoleucine	4.7
Threonine	6.1	Leucine	6.5
Serine	6.1	Tryptophan	–
Glutamic acid	9.2	Tyrosine	3.6
Proline	6.0	Phenylalanine	5.0
Glycine	9.8		

[a] From Ref. (27).

The purified protein has no inherent taste.   The effect from a 2.3 $\times 10^{-6}$ M solution of the protein may persist for over 3 hr.

Kurihara and Beidler (28) also studied the mode of action of their taste-modifying protein.   In one set of experiments they found that no change in thresholds for salty, bitter, and sweet qualities was induced by the protein.   They also investigated the effect of gymnemic acid on the taste-modifying properties of the protein.   First the mouth was rinsed with a solution of the protein.   At this point a 0.02 M citric acid solution, normally very sour, tasted sweet.   The mouth was then rinsed with a solution of gymnemic acid.   The same solution of citric acid now tasted sour.   This suggests that the protein does not depress the sourness, but rather that the taste modification is brought about by the addition of a sweet taste to the sour taste.   The authors suggested a mechanism for the protein action by postulating that "the protein binds closely to the vicinity of the sweet receptor site on the taste receptor membrane and a conformational change of the receptor membrane induced by sour substances (acids) results in 'fitting' the sugar residues of the taste-modifying protein to the sweet receptor site" (28).

It seems reasonable to expect that other substances possess taste-modifying properties.   If such compounds are isolated and identified, a better understanding may be gained of how our sense of taste functions.

III.   ODOR

Information concerning the chemical nature and specificity of odor receptors is even more limited than our knowledge of taste receptors. The problem is more complex.   Olfactory cells are much smaller and far more numerous than taste cells; the electrophysiologist is hampered in his attempts to record from single cells because of this small size. Gesteland et al. (29) have, however, recorded from single olfactory neurons in frogs, and their findings as well as results from grosser electrophysiological measurements again point to the nonspecificity of receptors.

A.   Structure and Odor

The relationship between structure and odor is even more elusive than the relationship between structure and taste. We assume, perhaps naively, that only four basic taste qualities exist, but the number of odors that we can distinguish may run into the thousands. Similar structures may possess markedly different odors, and very different structures may possess markedly similar odors.

Excellent examples of compounds possessing different structures and similar odors are the "musks." Theimer and Davies (30) in a review of musk-like odors and molecular structure included among the musks tricyclic compounds, macrocyclic ketones and lactones, indanes, tetralins, chromans, steroids, aromatic nitro compounds, etc. Structures of musk-like compounds are shown in Fig. 1-3.

Conversely, compounds whose structures appear to be similar may have quite different aromas. Stereoisomers, for example, may exhibit different odors. Naves (31) cites a number of examples. Thus, cis-3-hexenol has a characteristic fresh green odor while the trans-isomer has an odor reminiscent of chrysanthemum; of the four possible menthol isomers only menthol itself has the odor of peppermint. The iso-, neo-, and neoisomenthols all have disagreeable musty aromas. Even the lengthening of a carbon chain can markedly affect odor. γ-Nonalactone has a strong coconut-like aroma, while the γ-undecalactone is used to impart a peach aroma (15). Again the list of such compounds can be extended considerably.

B.   Odor Theories

The relationship between structure and odor is obviously elusive. Odor theories based on molecular size and shape, adsorption on a membrane surface, infrared radiation and adsorption, etc., have

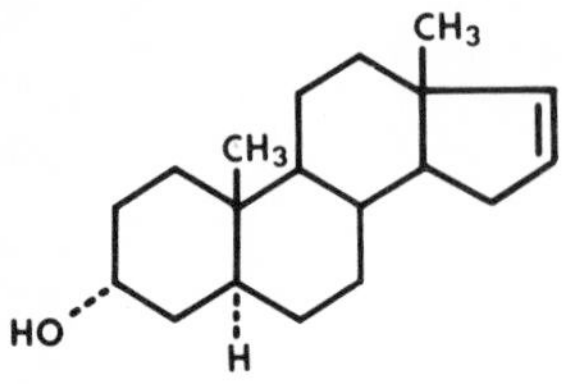

Δ^{16}-Androsten-3-α-ol

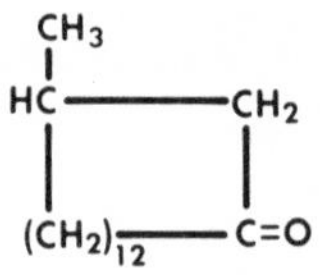

Muscone

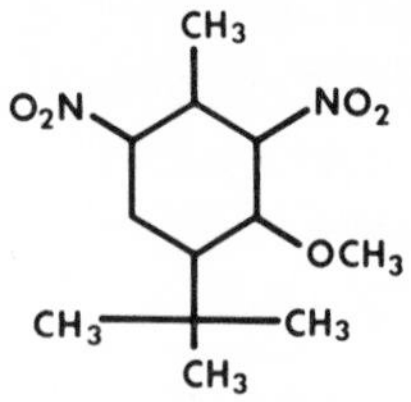

Musk ambrette

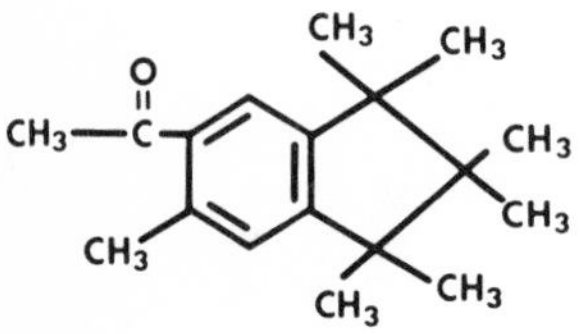

6-Acetyl-1,1,2,2,3,3,5-heptamethylindane

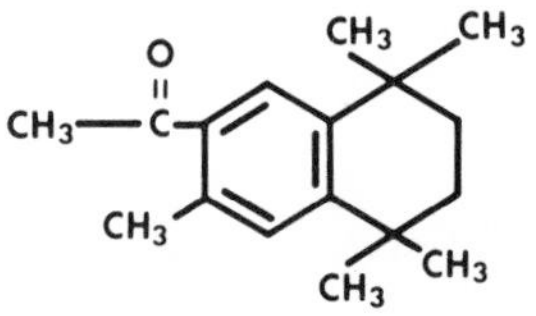

7-Acetyl-1,1,4,4,6-pentamethyl tetrahydronaphthalene

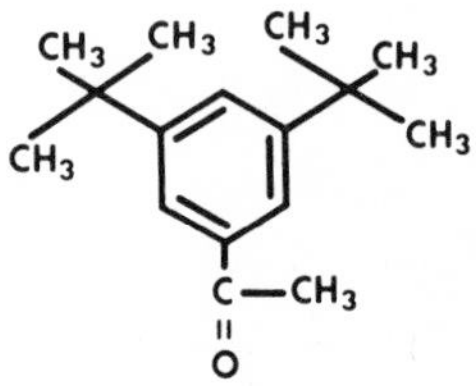

3,5-Di-tert-butylacetophenone

FIG. 1-3 Some musk-like compounds.

recently been reviewed by Dravnieks (32). The most encompassing theory has been developed by Davies and Taylor (33, 34) and Davies (35, 36). Their ideas provide (a) a triggering mechanism for initiating the odor response, (b) an explanation for the wide range in olfactory thresholds, (c) a means for calculating olfactory thresholds, and (d) a relationship between odor quality and physical properties.

The triggering mechanism for initiating the odor response is reminiscent of Beidler's work (10) and involves the adsorption of an odor molecule on a receptor site. This adsorption is sufficient to cause membrane "puncture." The resulting ion flow initiates the nerve impulse.

To explain the tremendous range in olfactory thresholds, which can vary from 10^8 to 10^{16} molecules of an odorant per milliliter of air, Davies and Taylor (34) postulate that, for a strong odorant, the adsorption of one molecule on a receptor site is sufficient to cause membrane "puncture" and initiate an odor response. For weaker odorants, two or perhaps three odorant molecules must be adsorbed per site before the membrane is "punctured." It follows that if two or more molecules must be adsorbed on the same small area before ion leakage across the membrane occurs, the concentration of odorant in the vapor phase must be much greater than when only one molecule is required, particularly since there are about 10-20 million olfactory receptors located in about 10 cm^2 of olfactory epithelium!

Two steps are involved for molecules of a given species within the nose to reach a receptor: the partitioning of molecules between air and the nasal mucosa (an aqueous phase), followed by a partitioning of the odorant molecues between the mucosa and the olfactory site (a lipid phase). These considerations plus calculations of the probability of finding odorant molecules within an area corresponding to that of a receptor site were used to derive an equation for calculating olfactory thresholds.

The measurements that must be made include:

a.   The cross-sectional area of an odorant molecule; this can be calculated from models or available bond distance.

b.   The partial pressure of the pure odorant in water, a measure of the partition coefficient between air and the nasal mucosa.

c.   The partition coefficient for the odorant compound between water and petroleum ether, a measure of the partition coefficient between the nasal mucosa and the receptor site.

Davies (36) has plotted the molecular cross section of odorant molecules versus their free energy of desorption at the lipid aqueous interface. This may be considered roughly equivalent to plotting the molecular shape versus a summation of the functional groups in the molecule. Davies has found that certain regions of the plot are associated with compounds of similar odors. These regions blend into one another. The implications are that a continuous spectrum of odors exists and that compounds of different structure may have similar odors if their shapes are roughly similar.

One interesting aspect of Davies' approach is that a rigid classification scheme is avoided. All odorous compounds fall within this system and their place in the odor spectrum is related to measurable quantities, i.e., their cross-sectional area, their vapor pressure, and the partition coefficient for the compound between an aqueous and an organic phase.

Olfaction theories are generally designed to explain odor quality in terms of so-called primary odors. Many of these theories are based on the assumption that odorous molecules possess steric forms which permit them to fit into suitable sites on the receptors that correspond to their own shape and size. In recent years a major exponent of this concept has been Amoore (37-39). He originally postulated seven primary odors but now thinks that rigid primaries are best replaced by broader "odor classes." These two theories by no means exhaust the ideas that have been advanced to explain odor quality. Radiation and vibration theories which imply action at a distance (i.e., radiation is emitted by the odorant molecule which is adsorbed by the receptor) have been proposed principally by Dyson (40,41) and by Wright (42,43), and enzyme theories in which odorant molecules are postulated to interact with enzymes in the receptors have also been proposed (44,45).

IV.   CHEMICAL ANALYSES

At this stage of theory development, the chemist cannot translate the proposed concepts into tools that can help unravel a flavor problem.

In addition to the absence of useful guidelines, there are other inherent difficulties in studying flavor that make this field a most challenging research area.

A. Unique Aspects of Flavor Research

Man's extreme sensitivity in detecting odors. Stuiver (46) has calculated that for the triggering of one human olfactory neuron by a powerful odorant as few as eight molecules are required, and that as few as 40 molecules can produce an identifiable olfactory sensation. If we assume that only one of 1000 molecules that are inspired reaches a receptor site, the other 999 being adsorbed or never diffusing into the olfactory region, then 40,000 molecules or about 10^{-19} moles can, at least theoretically, be detected by the nose, a sensitivity beyond the reach of our analytical techniques.

The diversity of compounds that may have an odor impact. One cannot limit the search for odorous compounds to, for example, esters or ketones. Any class of compounds may contain important odor contributors, and furthermore the food substrate may be equally complicated. It is very much like looking for a particular needle in a haystack full of needles!

Lack of objective sensory evaluation methods. Once a volatile compound derived from a given food has been isolated and identified, its contribution to the overall flavor picture must be assessed by subjective methods. A further complication is that the total flavor environment affects the contribution of the identified compound. Its contribution to flavor may be enhanced, inhibited, or modified by neighboring molecules.

Variation of olfactory thresholds for different compounds. Minor trace components may have far more sensory importance than other volatile compounds present in overwhelmingly greater amounts.

B. Basic Problems in Starting a Flavor Study

Sample preparation. The research worker will literally spend months studying a given sample or concentrate. Obviously, if the odor of this sample is not characteristic of the food, the results will be misleading. The sample must not only be representative but also reasonably stable. If the odor changes rapidly with time, artifacts may be identified.

Obtaining the sample represents only a small portion of the total effort expended in the investigation of a natural flavor, but it is an all-important step. Furthermore, if large quantities of foods must be processed to obtain a flavor concentrate adequate for identification of components, repetition of this step may be difficult. This problem may be particularly severe in investigations of some fruit and vegetable flavors for which large quantities of material must be processed and the fresh crop is available for only a few days or weeks each year.

Separation of complex mixtures. Having prepared a sample that merits analysis, how do we separate these complex mixtures into their individual components?

Identification of isolated compounds. Having separated the components, how do we identify them?

Sensory evaluation. How do we evaluate the sensory importance of the compounds isolated and identified?

Procedures that have been developed to solve these problems are briefly reviewed in this chapter. Subsequent chapters discuss these problems in depth.

C. Sample Preparation

Two distinct approaches and a combination of the two are available for isolating and/or concentrating the volatile compounds present in foodstuffs.

1. Total Volatile Analysis

In this approach one starts with large amounts of the food and by a judicious choice of distillation and extraction procedures finally obtains a flavor concentrate, hopefully free of artifacts, that contains all the volatile compounds that are present in the food.

At first glance one might suppose that a complete qualitative and quantitative analysis of these volatiles would enable one to reconstitute the original flavor by incorporating the pure compounds in the proper proportions into an appropriate substrate. This is an oversimplification. When one bites into an apple, the volatile mixture that reaches the olfactory epithelium differs quantitatively and perhaps qualitatively from the total volatile composition. The "volatility" of each of the volatile compounds present is roughly a function of its vapor pressure at body temperature and its degree of interaction with the food substrate. The composition of the volatile mixture in the mouth is of necessity different from the composition of the total volatiles. The total volatile composition tells us what compounds are present in the food, and we may gain some insight into their importance by observing their odors.

To reconstitute the original odor would depend, however, on the artistry of a flavorist. An interesting example is reported by Anderson and Day (47). Considerable qualitative and quantitative data had been obtained for blue cheese flavor. These compounds, in the amounts isolated, were added to a substrate composed of bland cottage cheese curd, milk fat, sweet cream, etc., and allowed to age for 12 hr. The resulting sample had a very intense soapy, bitter flavor. By changing the amounts of ketones and alcohols added and even eliminating some of the fatty acids, a final mixture closely resembling the natural flavor was developed.

Table 1-2 lists the compounds and concentrations used in preparing synthetic blue cheese flavor and compares these concentrations with the actual analytical data.

TABLE 1-2

A Comparison of the Composition of Synthetic and
Natural Blue Cheese Flavor[a]

Compound	Synthetic, mg/kg	Natural, mg/kg
Acetic acid	550	826
Butanoic acid	964	1448
Hexanoic acid	606	909
Octanoic acid	514	771
Acetone	6.2	3.1
2-Pentanone	30.3	15.2
2-Heptanone	69.5	34.8
2-Nonanone	66.3	33.1
2-Undecanone	17.0	8.5
2-Pentanol	0.9	0.4
2-Heptanol	12.1	6.1
2-Nonanol	7.0	3.5
2-Phenyl ethanol	2.0	---
Ethyl butanoate	1.5	---
Methyl hexanoate	6.0	---

[a]From Ref. (47).

2. Head Space Volatiles

It is conceivable for the total volatiles of two different foods to be
identical and for the two foods to have different odors. For example,
if the two foods vary in fat content, the vapor composition above
one will differ from the vapor composition over the other.

There are several advantages to looking at the head space volatiles directly:

(a) A relatively small sample of foodstuff is required.

(b) Very little sample preparation is needed; artifacts are therefore kept to a minimum.

(c) The compounds in the head space are present in the concentrations one actually smells.

One major disadvantage is that head space analysis pushes the available analytical techniques to their limits of sensitivity; thus, important high boiling flavor contributors that might be present in trace amounts may be assayed by the nose but missed by the analytical techniques.

A combination of total volatile analysis and head space analysis can be most useful. The total volatile analysis indicates the compounds that are present and the head space analysis helps to pinpoint their importance.

3. Concentration of Head Space Volatiles

Limited concentration of head space volatiles can be extremely useful. If head space volatiles are directly analyzed by gas chromatography, one may inject 20 ml of a gas sample. However, if a purified nitrogen gas stream is used to flush the head space volatiles into some kind of a collection system, then a 50-fold concentration can easily be obtained. Trace components previously undetectable can now be observed.

A simple system for the concentration of head space volatiles employs a trapping coil cooled in liquid nitrogen (48). The trap is essentially a continuation of a gas chromatographic column. The nitrogen sweep gas plus volatiles flow through the coil where the volatiles are frozen out. Modifications of this system have been used to collect volatile compounds from a variety of substrates (49, 50).

D. Isolation and Concentration of Total Volatiles

1. Distillation

In general, distillation of some kind is the procedure of choice for the isolation of all volatile materials from a food. Other methods such as extraction, adsorption, etc., are sometimes applied as a first step. If nondistillation procedures are used, volatiles as well as nonvolatiles are isolated and distillation must be applied at a later stage. The methods other than distillation can be considered preliminary concentration steps.

a. Simple Distillation. This can be carried out at atmospheric pressure or reduced pressure. An extremely useful technique is high vacuum degassing and stripping. These methods have been extremely useful in removing volatiles from fats, oils, and high boiling solvents (51,52).

b. Steam Distillation. This technique is most useful for removal of higher boiling compounds that can be volatilized with steam either at atmospheric or somewhat reduced pressure.

c. Fractional Distillation. This too can be conducted at atmospheric pressure or under vacuum. At this point of sample preparation fractional distillation serves to concentrate the volatiles by removal of water by "reflux stripping."

Specific rules for selecting the best distillation procedure cannot be given. Each problem must be evaluated individually. In general, vacuum distillation tends to diminish the chances of thermal degradation. Steam distillation also minimizes the danger of overheating since it can be continued until all odor volatiles are removed without changing the volume of liquid in the stillpot.

Some common-sense precautions should also be observed. (1) If one's interest is in studying the aroma of fresh uncooked foods, distillation at atmospheric pressure should be avoided. (2) Every possible precaution must be exercised to ensure that the isolation procedure does not produce artifacts. Think of one's chagrin if a compound derived from solvents, tubing, stopcock grease, etc. is isolated and identified as a flavor contributor. (3) To check if heat-induced changes have taken place one may observe after each step if the aroma has appreciably changed.

2. Concentration

When a distillate containing the volatiles has been obtained, particularly if steam distillation has been used, one may have a dilute aqueous solution that must be concentrated.

a. Extraction. If the solution is not too dilute, extraction with organic solvents can be made directly. For some mixtures continuous extraction is useful; e.g., continuous extraction with pentane is often used to extract organic material from an aqueous solution containing comparatively large amounts of alcohol. The pentane extracts the organic compounds and leaves most of the alcohol in the aqueous phase. If interest is centered on low boiling solutes in the aqueous phase, Weurman (53) suggests not to use low boiling solvents for extraction. Low boiling volatiles will escape when the solvent is distilled. It may be preferable to use a high boiling solvent and strip the solvent at reduced pressure.

b. Freeze Concentration. If the solution is extremely dilute one may require a concentration step prior to extraction. Freezing can readily give an 10- to 20-fold concentration of volatile materials with about an 80% recovery of low boiling volatiles (54). To do this efficiently the mixture of ice and water must be continually stirred and a portion of the solution must always remain unfrozen. A recent example of freeze concentration of a steam distillate is reported by van Praag et al. (55) in a study on cocoa flavor.

c. Charcoal Adsorption. This technique may also be used to
concentrate very dilute solutions. Charcoal is not deactivated by
water and has a great capacity for adsorbing organic compounds. In
fact, the more dilute the aqueous solution being passed through the
adsorbent, the more efficient the process. Desorption from charcoal
can be accomplished by elution with solvents such as ethyl ether or
carbon disulfide; e. g. , Carson and Wong (56) in a study of onion
volatiles adsorbed the volatiles on charcoal and eluted them by Soxhlet
extraction with ether. Preliminary freeze drying of the charcoal to
remove water prior to elution with organic solvents can also be used
(57).

E. Separation of Mixtures

Gas chromatography (GC) is the method of choice for separating
small amounts of complex mixtures and is versatile enough to handle
head space volatiles or flavor concentrates.

In some instances, because of the complexity of the concentrate,
gross fractionation by chemical procedures can simplify subsequent
GC separations. For example, a preliminary acid-base extraction
may separate a mixture into acidic, basic, and neutral fractions. The
fractions can then be separated by GC.

Recently, Murray and Stanley (58) described a microcolumn technique
for the preliminary separation of 5-50 mg of flavor mixtures into
classes based on polarity. The method, originally devised for the
separation of unsaturated alcohols from paraffin hydrocarbons in peas,
should have wide application in other flavor studies.

1. Column Selection

For the GC separations of complex mixtures (even after preliminary
fractionation) a variety of columns and various stationary liquid phases
with a range of polarity and selectivity are needed. No infallible rules
can be promulgated to predict which column will separate what mixture. Resolution in GC depends on the differences, under identical

conditions of chromatography, among the partition coefficients of the compounds being separated and on the efficiency of the column. Compounds that cannot be well resolved on one phase may be separable on another phase of different polarity.

The efficiency of a chromatographic column is a measure of peak broadening as the solutes pass through the column. The more efficient the column, the less broadening occurs. The greater the efficiency of the column, the better the resolution of compounds whose partition co-efficients may differ only slightly.

Highest efficiencies in GC separations are attained using 0.01-in. capillary columns. However, such columns have a very limited capacity and can easily be ruined by overloading. Larger diameter capillaries have been developed that give a reasonable compromise between efficiency and capacity (59). The 0.02-in. capillary is easier to handle than the 0.01-in. capillary and offers a reasonable com-promise between efficiency and quantity of material that can be separated. The 0.03-in. column can handle quantities usually associ-ated with separation on a 1/8-in. packed column. Mon et al. (60), in a study on hop oil, compared results obtained using a packed versus a capillary column. The capillary gave higher resolution, but the packed column separated 1000 times more material. The packed column also required 20 times as long to complete the separation. Advantage can be taken of the capacity of packed columns for the initial separation of large samples. The final resolution of the collected fractions can be made on capillaries.

2. Temperature Programming

Because of the wide range in the boiling points of flavor volatiles, temperature programming is almost always used. Subambient programming has been used by Merritt and Walsh (61) for the separation of low boiling compounds. Programming can start below -100°C and can continue up to +100°C or higher, depending on the nature of the liquid phase. Lowering the initial temperature in programmed GC

improves the separation of the lower boiling components.   The
separation of higher boiling materials is essentially unchanged.

Generally, a preliminary GC separation on a nonpolar column
should be made and fractions collected between arbitrarily selected
temperatures.   These fractions should be evaluated organoleptically
to establish their contribution to the flavor under study.   Maximum
effort can then be expended on the fractions that contain the most
characteristic odorants.

F. Identification of Components

Some information can be gained from relative retention data.   The
Kovats retention index (62) is widely used and expresses the retention
behavior of an unknown using the normal paraffins as standards.   In
general, retention indices are most useful for identifying monofunctional
compounds of low molecular weight.   When larger molecules of
multiple functionality are isolated, identification based solely on
retention data is rather questionable.

To determine the structure of a compound unambiguously, it is
generally necessary to use additional means of identification.   Compounds emerging from the gas chromatograph can be looked at "on the
fly, " providing we can obtain some kind of a useful record in a matter
of seconds.   Alternatively, each compound can be trapped as it is
eluted and examined at one's leisure.   Trapping systems can be very
simple — a melting point tube inserted in the GC outlet is surprisingly
efficient for collecting high boiling compounds that might form aerosols.
Basically, materials condense when cooled below their dew point.   If
cooled too quickly, too many nuclei form and an aerosol results.
Cooling along a temperature gradient results in maximum condensation.
The efficiency of a glass capillary in trapping high boiling solutes is
a result of the temperature gradient along the tube from the GC outlet
to room temperature.   For very volatile materials, the effluent gas
can be bubbled through a small amount of cold carbon tetrachloride.

The organic solutes will be dissolved in the solvent. The resulting solution is particularly suitable for IR or NMR studies. Other systems that have been used include the following: trapping on KBr crystals for use with IR spectroscopy (63), total trapping of carrier gas plus solute (64, 65), traps designed for transferring fractions to another GC column or to a mass spectrometer (66).

Once a compound has been isolated in a pure state its structure must be determined. The most powerful methods are the spectrometric procedures.

Infrared spectroscopy provides one of the best means of identifying functional groups as well as "fingerprinting" the molecule. Usually about 100 μg of a compound are required. For some compounds, and with instruments equipped with scale expansion, beam condensers, etc., samples in the 1- to 10-μg range can give useful information. Raman spectrometers using laser sources are also being used to study flavor volatiles. Again, about 100 μg is the current minimum level of detection.

Nuclear magnetic resonance is a potent tool for elucidating structure but lack of sensitivity has limited its use in flavor studies. However, practical microcells, time averaging, and recently introduced pulsed NMR improve sensitivity.

Mass spectrometry (MS), despite its high cost, is widely used for studying flavor volatiles because of one major advantage: The amount of material required to obtain a spectrum is in the same range as the amount that can be measured by GC. One μg or less can give a useful spectrum; practical sensitivity limits are determined by column bleed and spectrometer background.

The mass spectrometer can be used in two ways. Samples can be trapped and examined at leisure. This can create a major storage problem; e.g., more than 350 compounds have been identified in coffee! Alternatively, a fast scan mass spectrometer and gas

chromatograph can be combined in tandem into one instrument. This
combination may be the instrument of choice.

The hookup between the GC and MS is critical. The problem is that
the pressure in the ion source of the mass spectrometer is about 10^{-6}
Torr and the carrier gas is at atmospheric pressure as it emerges
from the GC. Carrier gas, usually helium, must therefore be
removed preferentially from the carrier gas-solute combination.
Separators for this purpose have been developed (67-69). Improvements in the methods for coupling the GC-MS would be highly desirable.

In addition to the spectrometric techniques, reactions that can be
carried out in some kind of a precolumn or on the GC column can be
made to yield useful analytical information. Hydrogenation is a case
in point. Hydrogen can be used as the carrier gas and, with the
proper catalyst, temperature, and time, three controlled vapor phase
reactions may take place: hydrogenation (addition of hydrogen),
hydrogenolysis (cleavage with addition of hydrogen at cleavage point),
and dehydrogenation (hydrogen abstraction). Hydrogenation and
hydrogenolysis have been most widely used. Examples of the value
of these techniques have been described by Beroza and Sarmiento
(70, 71) and others (72, 73).

There are a variety of synthetic and degradative reactions which
can be applied on a microgram scale yielding reaction products which
can then be examined by GC. For example, double-bond position, in
many instances, can be determined by ozonolysis and subsequent GC
of the reaction products (74).

Chemical methods consisting of specific tests for functional groups
in compounds that are emerging from the GC column have been
described by Walsh and Merritt (75). A subtractive technique in which
components of a mixture are reacted with reagents coating the walls of
the injection syringe has been described by Hoff and Feit (76); the
chromatograms of the original mixture and the reacted mixture are
then compared.

It is clear that no protocol for identifying a compound can be offered. The method selected for the identification of a purified compound will vary with the complexity of the molecule, the instrumentation available, and the investigator's background and training.

V. SENSORY EVALUATION

Isolation, identification, and quantitative determination of volatile food components is clearly a complex problem. Evaluation of the contribution of these compounds to food flavor is undoubtedly as difficult. Many methods, including threshold measurements, difference tests, and descriptive tests, have been used. One can reasonably assume that if a substance is present above its threshold it has significance, but sensory thresholds are not invariant. These values can be influenced by internal and external "noise" (i.e., the conditions of testing and one's own well-being). Interpretation of threshold data is further complicated by the fact that some compounds modify an odor even if they are present below threshold values. Additive effects also take place; mixtures of two or more components each below threshold may produce a detectable odor.

One interesting approach to the evaluation of the chromatographic data has been taken by Biggers et al. (77), who have attempted to determine the peaks in a chromatogram that can serve as quality indices to coffee flavor. A computer program was prepared by which the ratio of every peak in a chromatogram to every other peak in the same chromatogram could be determined. These ratios for one coffee were compared to similar ratios for another coffee. By doing this with coffees of different organoleptic quality, discriminating ratios that permitted classification of coffees as "good" or "bad" were obtained. The situation is complicated because the chromatogram of a given coffee will vary with the degree of roast. Initially, discriminating ratios valid for coffee varieties roasted to optimum color were determined and then those ratios which were meaningful, independent of roast, were obtained.

In a study of odor intensities, Guadagni et al. (78) employed the concept of the odor unit, Uo, introduced by Rothe and Thomas (79). One odor unit is the olfactory threshold concentration, Tc, for a given compound in parts per billion. The ratio in a given sample of the actual concentration of a given compound to the threshold concentration of that compound represents the number of odor units contributed by that component; e.g., if Tc = 1 ppb and the compound is present at 10 ppb, then this component contributes 10 odor units. The assumption is then made that odor units are additive. One value of this approach is that it enables one to estimate the contribution of a compound to the overall aroma in terms of odor rather than in terms of concentration.

VI. SUMMARY

The state of the art of flavor research and the problems and pitfalls facing the chemist as he approaches a problem in flavor chemistry have been discussed. The approaches, techniques, and methodology that can be utilized have been outlined. The following chapters will supply the muscle for the skeleton that has been assembled.

REFERENCES

1. R. L. Hall, Food Technol., 22, 1388 (1968).

2. J. A. Brillat-Savarin, The Physiology of Taste (transl.), Dover, New York, 1960, p. 26.

3. R. W. Moncrieff, The Chemical Senses, L. Hill, London, 1944, p. 168.

4. C. Pfaffman, J. Neurophysiol., 18, 429 (1955).

5. K. Kimura and L. M. Beidler, J. Cellular Comp. Physiol., 58, 131 (1961).

6. F. R. Dastoli and S. Price, Science, 154, 905 (1966).

7. Y. Hiji, N. Kobayashi, and M. Sato, Kumamoto Med. J., 21, 137 (1968).

8. F. R. Dastoli, D. V. Lopiekes, and A. R. Doig, Nature, 218, 884 (1968).

9. S. Price, J. Agr. Food Chem., 17, 709 (1969).

10. L. M. Beidler, J. Gen. Physiol., 38, 133 (1954).

11. L. M. Beidler, Progress in Biophysics and Biophysical Chemistry, Pergamon, London, 1961, p. 107.

12. R. S. Shallenberger, J. Food Sci., 28, 584 (1963).

13. R. S. Shallenberger and T. E. Acree, J. Agr. Food Chem., 17, 701 (1969).

14. J. Solms, L. Vuataz, and R. H. Egli, Experientia, 21, 692 (1965).

15. K. Kulka, J. Agr. Food Chem., 15, 48 (1967).

16. H. Staudinger and H. Schneider, Chem. Ber., 56B, 699 (1923).

17. T. Takemoto and T. Nakajima, Yakugaku Zasshi, 84, 1183 (1964).

18. A. Kuninaka, in Flavor Chemistry, Advances in Chemistry Series (R. F. Gould, ed.), American Chemical Society, Washington, D. C. 1966, Chap. 15.

19. Y. Kawamura, A. Adachi, M. Ohara, and S. Ikeda, Symp. Amino Acid Nucleic Acid, 10, 168 (1964).

20. D. Hooper, Pharm. J. Trans., 17, 867 (1887).

21. R. M. Warren and C. Pfaffman, J. Appl. Physiol., 14, 40 (1959).

22. W. Stöcklin, E. Weiss, and T. Reichstein, Helv. Chim. Acta, 50, 474 (1967).

23. W. Stöcklin, J. Agr. Food Chem., 17, 704 (1969).

24. Y. Kurihara, Life Sci., 8, 537 (1969).

25. W. F. Daniell, Pharm. J., 11, 445 (1852).

26. G. E. Inglett, B. Dowling, J. J. Albrecht, and F. A. Hoglan, J. Agr. Food Chem., 13, 284 (1965).

27. K. Kurihara and L. M. Beidler, Science, 161, 1241 (1968).

28. K. Kurihara and L. M. Beidler, Nature, 222, 1176 (1969).

29. R. C. Gesteland, J. Y. Lettvin, W. H. Pitts, and A. Rojas, in Proc. 1st Intern. Symp. Olfaction and Taste (Y. Zotterman, ed.), Vol. 1, Pergamon, New York 1963, pp. 19-34.

30. E. T. Theimer and J. T. Davies, J. Agr. Food Chem., 15, 6 (1967).

31. Y. R. Naves, in Molecular Structure and Organoleptic Quality, Society of Chemical Industry, London, 1957, pp. 38-53.

32. A. Dravnieks, in Flavor Chemistry, Advances in Chemistry Series (R. F. Gould, ed.), American Chemical Society, Washington, D. C., 1966, pp 29-52.

33. J. T. Davies and F. H. Taylor, Nature, 174, 693 (1954).

34. J. T. Davies and F. H. Taylor, Biol. Bull. Woods Hole, 117, 222 (1959).

35. J. T. Davies, Symp. Soc. Exptl. Biol., 16, 170 (1962).

36. J. T. Davies, J. Theoret. Biol., 8, 1 (1965).

37. J. E. Amoore, Perfumery Essent. Oil Record, 43, 321 (1952).

38. J. E. Amoore, Proc. Sci. Sect. Toilet Goods Assoc., Suppl. 37, 1 (1962).

39. J. E. Amoore, Ann. N. Y. Acad. Sci., 116, 457 (1964).

40. G. M. Dyson, Perfumery Essent. Oil Record, 28, 13 (1937).

41. G. M. Dyson, Chem. Ind. (London), 16, 647 (1938).

42. R. H. Wright, J. Appl. Chem., 4, 611 (1954).

43. R. H. Wright, Science of Smell, Basic Books, New York, 1964.

44. G. B. Kistiakowski, Science, 112, 154 (1950).

45. L. Ruzicka, in Molecular Structure and Organoleptic Quality, Society of Chemical Industry, London, 1957, pp. 116-124.

46. M. Stuiver, Doctoral Thesis, Rijks University, Groningen, Holland, 1958: through H. Devries and M. Stuiver, in Sensory Communication (W. A. Rosenblith, ed.), Wiley, New York, 1961, pp. 159-167.

47. D. F. Anderson and E. A. Day, J. Agr. Food Chem., 14, 241 (1966).

48. I. Hornstein and P. F. Crowe, Anal. Chem., 34, 1354 (1962).

49. M. E. Morgan and E. A. Day, J. Dairy Sci., 48, 1382 (1965).

50. R. G. Arnold and R. C. Lindsay, J. Dairy Sci., 51, 224 (1968).

51. C. H. Lea and P. A. T. Swoboda, J. Sci. Food Agr., 13, 148 (1962).

52. D. A. Forss, V. M. Jacobsen, and E. H. Ramshaw, J. Agr. Food Chem., 15, 1104 (1967).

53. C. Weurman, J. Agr. Food Chem., 17, 370 (1969).

54. J. Shapiro, Anal. Chem., 39, 280 (1967).

55. M. van Praag, H. S. Stein, and M. S. Tibbets, J. Agr. Food Chem., 16, 1005 (1968).

56. J. F. Carson and F. F. Wong, J. Agr. Food Chem., 9, 140 (1961).

57. D. E. Heinz, M. R. Sevenants, and W. G. Jennings, J. Food Sci., 31, 69 (1966).

58. K. E. Murray and G. Stanley, J. Chromatog., 34, 174 (1968).

59. R. Teranishi and T. R. Mon, Anal. Chem., 36, 1490 (1964).

60. T. R. Mon, R. A. Flath, R. R. Forrey, and R. Teranishi, in Advances in Gas Chromatography (A. Zlatkis, ed.), Preston Technical Abstracts, Evanston, Ill., 1967, pp. 30-33.

61. C. Merritt, Jr., and J. T. Walsh, Anal. Chem., 35, 110 (1963).

62. E. sz. Kovats, in Advances in Chromatography (J. C. Giddings and R. A. Keller, eds.), Vol. 1, Dekker, New York, 1965, pp. 229-247.

63. L. Guiffrida, J. Assoc. Offic. Agr. Chem., 48, 354 (1965).

64. I. Hornstein and P. Crowe, Anal. Chem., 37, 170 (1965).

65. P. A. T. Swoboda, Nature, 199, 31 (1963).

66. M. L. Bazinet and J. T. Walsh, Rev. Sci. Instr., 31, 346 (1960).

67. J. T. Watson and K. Biemann, Anal. Chem., 37, 845 (1965).

68. R. Ryhage, Arkiv Kemi, 26, 305 (1967).

69. P. M. Llewellyn and D. P. Littlejohn, paper presented at Pittsburgh Conference on Analytical Chemistry and Applied Spectrometry, February, 1966.

70. M. Beroza and R. Sarmiento, Anal. Chem., 35, 1353 (1963).

71. M. Beroza and R. Sarmiento, Anal. Chem., 38, 1042 (1966).

72. R. M. Teeter, C. F. Spencer, J. W. Green, and L. H. Smithson, J. Am. Oil Chem. Soc., 43, 82 (1966).

73. T. L. Mounts and H. T. Dutton, Anal. Chem., 37, 641 (1965).

74. M. Beroza and B. A. Bierl, Anal. Chem., 39, 1131 (1967).

75. J. T. Walsh and C. Merritt, Jr., Anal. Chem., 32, 1378 (1960).

76. J. E. Hoff and E. D. Feit, Anal. Chem., 36, 1002 (1964).

77. R. E. Biggers, J. J. Hilton, and M. A. Gianturco, in Advances in Gas Chromatography (A. Zlatkis, ed.), Preston Technical Abstracts, Evanston, Ill., 1969, pp. 236-255.

78. D. G. Guadagni, S. Okano, R. G. Buttery, and H. K. Burr, Food Technol., 20, 166 (1966).

79. M. Rothe and B. Thomas, Z. Lebensm. Untersuch. Forsch., 119, 302 (1963).

RELATED GENERAL REFERENCES

1. M. A. Amerine, R. M. Pangborn, and E. B. Roessler, Principles of Sensory Evaluation of Food, Academic, New York, 1965.

2. R. M. Benjamin, B. P. Halpern, D. G. Moulton, and M. M. Mozell, Ann. Rev. Physiol., 16, 381 (1965).

3. R. W. Moncrieff, The Chemical Senses, 3rd ed., CRC Press, Cleveland, 1967.

4. M. R. Kare and B. P. Halpern, The Physiological and Behavioral Aspects of Taste, Univ. Chicago Press, 1961.

5. W. A. Rosenblith, Sensory Communication, Mass. Inst. Technol. Press and Wiley, Cambridge, 1961.

6. Y. Zotterman, Olfaction and Taste, Proc. 1st Intern. Symp., Stockholm, 1962, Pergamon, New York, 1963.

7. T. Hayashi, Olfaction and Taste, Proc. 2nd Intern. Symp., Tokyo, 1965, Pergamon, New York, 1967.

8. Basic Principles of Sensory Evaluation, ASTM Committee E-18 on Sensory Evaluation of Materials and Products, STP 433, American Society for Testing and Materials, Philadelphia, 1968.

9. Manual on Sensory Testing Methods, ASTM Committee E-18 on Sensory Evaluation of Materials and Products, STP 434, American Society for Testing and Materials, Philadelphia, 1968.

10. Correlation of Subjective-Objective Methods in the Study of Odors, ASTM Committee E-18 on Sensory Evaluation of Materials and Products, STP 440, American Society for Testing Materials, Philadelphia, 1968.

Chapter 2

ISOLATION OF FLAVOR CONCENTRATES

I. INTRODUCTION

Flavor has been defined as the sum of those characteristics of any "material taken in the mouth, perceived principally by the senses of taste and smell, and also by the general pain, tactile and temperature receptors in the mouth, as received and interpreted by the brain" (1). Though each of us might wish to modify this definition somewhat to meet our own views, it certainly makes clear the fascinating complexity of stimuli that give rise to the sensation we call flavor. The term "flavor" is widely used, but often not in the above general sense. Instead, one means the flavor contribution made by the volatile constituents of foods. To be exact, one primarily means aroma. Thus, this chapter might also be called "Isolation of Aroma Concentrates." However, because aroma makes a very large contribution to flavor, the terms "aroma concentrates" and "flavor concentrates" are used interchangeably.

Aromas have their origins in the nature and composition of the vapor over foods. Their qualitative and quantitative compositions

37

depend upon the vapor pressures of food constituents and on the extent of interaction of the volatile components with nonvolatile food components. Direct vapor analysis (head space analysis) of food aromas could, in theory, provide all necessary information about aroma composition, if gas chromatographic columns possessed ultimate resolving power and if the sensitivity and stimulus-response behavior of gas chromatographic detectors were the same as those of the nose. In reality, achieving understanding of an aroma often depends as much on knowledge of the identity and quantity of volatiles in the food itself as on knowledge of the head space vapor composition. In order to determine the identity and quantity of volatile food constituents, the isolation of aroma concentrates is very often necessary — hence, the detailed attention given to this subject.

Let us refresh our memories concerning the gross chemical composition of foods (Table 2-1). The ranges given are very broad in order to include a wide variety of foodstuffs, but they illustrate in a very real sense the order of magnitude of the quantities in which the

TABLE 2-1

Food Composition

Water	up to 95%
Protein	1-25%
Lipid	1-40%
Carbohydrates	1-80%
Minerals	1- 5%
Vitamins	ppm
Flavor and aroma compounds	ppm and ppb

gross constituents are found. The flavor and aroma compounds are present in extremely small amounts: parts per million if we are lucky, but often parts per billion! Therein lies the challenge.

Water is by far the major volatile constituent. It is isolated along with the other volatile constituents, which are also the most minor constituents, and it cannot be separated from them. What is meant by "volatile"? The definition is operational. It includes (1) compounds with sufficient vapor pressure to permit them to be in the air where we can smell them; but it also includes (2) compounds that are volatile under conditions of high vacuum distillation, since that is often a necessary procedure for obtaining an aroma concentrate from a food product. As we use the term, "volatile" covers compounds having a very wide range of boiling points — from noncondensable gases to substances that boil at temperatures above 300° C.

The task is to obtain from the food product of interest a quantity of the very complex mixture of its volatile components sufficiently large to allow their individual chemical identification. It is required that the resulting mixture or concentrate exhibit the characteristic odor of the foodstuff when reconstituted. If the biological activity (aroma) of the concentrate is not appropriate, there is no need to identify its components.

Distillation, extraction, or other procedures necessary for the isolation of adequate quantities of odor concentrates by their nature cause preferential accumulation of certain components and loss of others. During such procedures components are transferred from one medium to another — from the foodstuff to an aqueous or organic solvent in which they may be more or less soluble than in the food matrix. The original quantitative interrelationship of aroma components is destroyed, particularly in the vapor over the extract or distillate. In view of the compositional changes that are

necessarily brought about by isolation of flavor concentrates, it is remarkable that the concentrates can retain so well their original aroma
quality!

Because of the major disruption caused in foods by isolation of
aroma concentrates, selection of the isolation procedure is critically
important. Mistakes made at this stage cannot be remedied or overcome. Great care must be taken to avoid introduction of contaminants
or generation of artifacts. Our knowledge of flavor compounds is
directly dependent on the methods used for their isolation. There is
real need for innovation and creativity in the improvement of such
methods. A number of these are outlined in Table 2-2.

II. METHODS FOR ISOLATION AND CONCENTRATION OF VOLATILE FOOD CONSTITUENTS

For isolation of maximum quantities of volatiles, the food product
of interest should be taken at its stage of optimum aroma quality and
intensity. This might be an appropriate stage of ripeness or
maturity, or an optimum stage of processing, i.e., roasting, boiling,
etc. In certain cases the quantity of volatiles may be augmented by
grinding, chopping, pressing, or otherwise breaking the natural structure of the product. One must be careful, however, that the flavor is
not destroyed due to chemical or enzymatic change. Low temperatures, inert atmospheres, or, at the other extreme, "cooking" conditions may be required to generate appropriate aroma quality and
intensity.

Since, by definition, aroma is caused by compounds volatile
enough to reach our odor receptors, it is wise to try to remove the
volatiles from the nonvolatile food constituents. Distillation is thus
very often the first method of choice. For products other than fats
and oils, water is the major volatile component. Water is also, in a
sense, the major "contaminant" because it makes no contribution to
flavor or aroma and its removal often poses difficult problems.

TABLE 2-2

Methods for Isolation and Concentration
of Volatile Food Constituents

Distillation

Flash distillation
  Atmospheric pressure
  Reduced pressure

Steam distillation
  Atmospheric pressure
  Reduced pressure

CO_2 distillation

Vacuum fractional distillation

High vacuum distillation
  Degassing
  Molecular distillation

Other Types

Gas entrainment
  Open system
  Closed system

Distillation of aqueous and nonaqueous extracts

Lyophylization (freeze drying)
  Retention of volatiles

Freeze concentration
  Zone melting

Adsorption
  Charcoal
  Silica gel chromatography
  Dry column chromatography

Extraction

Derivative formation

However, its presence in many foods or sources of flavoring ingredients can be helpful because it makes removal of less volatile water-insoluble organic compounds possible through distillation.

Sometimes the starting material is already a concentrate in that it may be an essence, distilled spirits, or an essential oil. Separation from nonvolatile materials has already been accomplished or is not needed. Even in these cases, however, distillation may be selected as the most useful initial concentration procedure.

A. Distillation

1. Recovery of Volatiles from Aqueous Media

The design of distillation equipment for essence recovery in the production of high quality fruit and vegetable concentrates has been actively studied. Problems encountered in volatile flavor recovery are discussed by Walker (2). If the aroma is not damaged by high temperatures, flash evaporation at atmospheric pressure is feasible. In certain equipment this can be accomplished for heat-sensitive materials. Roger and Turkot (3) described an apparatus which included a "rapid atmospheric evaporator which strips the volatile aromas from heat-sensitive fruit juices without damage and a continuous-distillation column which rectifies the volatile flavoring compounds to 100-150 times their original concentration in the single-strength juice." For purposes of isolation and identification of aroma constituents, the aroma-containing distillate would be subjected to further concentration. In commercial procedures it would be added back to the concentrated juice. The authors (3) discuss losses that may occur with certain important flavor constituents; for example, methyl anthranilate in Concord grape juices. Losses are attributed to inefficient stripping in the column. Suggestions are made for improved stripping column design that can be expected to afford better recoveries of important aroma components. Factors involved in achieving flavor retention and heat transfer during concentration of liquids in film evaporators also have been evaluated (4).

An alternate method (5) is particularly designed for use with heat-sensitive aromas in that it permits the feed to a WURVAC column to be boiled under vacuum at 100° F (see Fig. 2-1) — milder conditions than those possible at atmospheric pressure. The major problem encountered in collection of essences under vacuum is loss of aroma compounds by entrainment in the noncondensable gases that enter with the feed and via small air leaks. This problem is overcome and, in fact, used to advantage in the apparatus shown in Fig. 2-1 (5). Feed is introduced to the evaporator at rates between 100 and 200 lb/hr. The vapors go to the stripping column and the stripped juice to the cooler and receivers. "The unique feature of this process is the manner in which volatiles are removed from the stripping column. A liquid-sealed vacuum pump with recirculating sealant provides a convenient way of absorbing volatiles leaving with the noncondensable gases. A vacuum bleed to the reboiler allows the volatile flavor to be recovered by means of these noncondensable gases" (5). The nitrogen bleed carries aroma from the condenser and into a water-sealed vacuum pump which compresses the gas to atmospheric pressure and mixes it thoroughly with the water sealant. The water is continually circulated through the pump to absorb vapors and through a cooler to maintain its temperature at 36° F. The effectiveness of this procedure is illustrated in Fig. 2-2, which shows gas chromatographically the vapor composition over feed (fresh apple juice), stripped feed, column bottoms, and the aroma solution. These results were confirmed by taste panel evaluations that rated the concentrates of high quality. In addition to several useful commercial applications, the WURVAC process provides a means of obtaining aroma concentrates of great value to flavor chemists. Apple, orange, and tomato juices and apricot and peach purees have been processed in this manner (5).

If aroma constituents can be steam-distilled, then a combination steam distillation-continuous extraction apparatus, described by

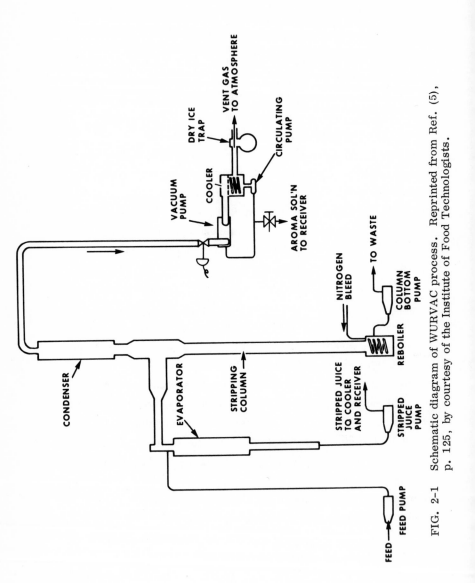

FIG. 2-1 Schematic diagram of WURVAC process. Reprinted from Ref. (5), p. 125, by courtesy of the Institute of Food Technologists.

44

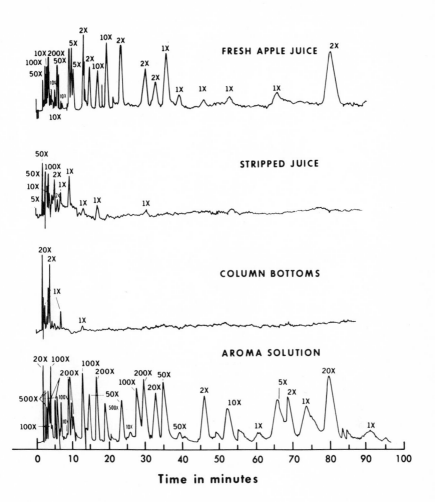

FIG. 2-2   Aromagrams of apple juice fractions. Reprinted
from Ref. (5), p. 126, by courtesy of the Institute
of Food Technologists.

Nickerson and Likens (6) (see Fig. 2-3), can be used. This apparatus
allows simultaneous condensation of the steam distillate and an
immiscible extracting solvent.   The distillates return to their respective distillation flasks via arms at different levels, e. g. , water phase

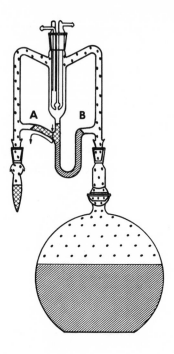

FIG. 2-3   Distillation-extraction Apparatus. Reprinted from
Ref. (6), p. 2, by courtesy of Elsevier Publishing
Company.

through arm B and pentane through arm A. This method has the very
desirable feature of using the solvent over and over. The minimum
solvent volume lessens the introduction of contaminants from the
solvent, lowers entrainment losses of the desirable volatiles in the
solvent distillation step, and permits a very high concentration to be
effected in one step.

It is, of course, very desirable when large volumes of aqueous
distillates can be made to yield flavor compounds without extraction
with organic solvents. Forss et al. (7) described two methods to
accomplish this. One employed distillation of 5- to 500-ml volumes
of aqueous solutions at 0-20 Torr through a cold (0° C) vertical condenser and collection of the resulting small volumes (1/1000 to 1/50

of the original) in a liquid nitrogen trap. About 400-fold concentration was achieved with yields of over 80% for 1-alkanols, 2-alkanones, and ethyl alkanoates boiling below 175°C. About 200-fold concentration was accomplished by high vacuum sublimation through a trap held at -55°C and collection of volatiles in a liquid nitrogen trap. This procedure yielded concentrates of "lower boiling compounds" almost free of water (7).

2. Recovery of Volatiles from Fats and Oils

Though steam distillation (8-11) and sweeping with an inert gas (12) have been used to remove volatiles from lipids, high vacuum distillation appears, on the whole, to be the most useful process. As with all distillation methods, quantitative results are very difficult to reproduce from batch to batch. Low molecular weight compounds (C_1 to C_8) tend to be lost due to evaporation before analysis or because they are not totally condensed in cold traps. With increasing molecular weight, lower recoveries of volatiles are achieved due to their increasing lipid solubility and their decreasing vapor pressure.

C_3 through C_8 methyl ketones were recovered (13) at 78-99% yield from butterfat after 8 hr of vacuum distillation at 45°C, but decreased yields were obtained from nonanone (87%) through dodecanone. High vacuum degassing of butter oil at 50°C resulted in less than 50% recovery of methyl ketones above C_8 or of alcohols above C_5. When degassing was followed by a cold-finger molecular distillation, improved recoveries were possible. Advantages and disadvantages of reduced pressure steam distillation, cold-finger molecular distillation, and high vacuum degassing for isolation of volatiles from butter oil were compared (13).

Recently Nawar et al. (14) demonstrated excellent recovery of alkanes and alkenes from corn oil with an apparatus based on that of deBruyn and Schogt (15). C_2 through C_{10} hydrocarbons were essentially completely recovered at 10^{-3} Torr and 70°C by collection in a precolumn (P) packed with stainless steel helices and fitted on both

sides with high vacuum temperature-resistant valves (V). The
apparatus is shown in Fig. 2-4. The C_{11} through C_{22} hydrocarbons
were removed from corn oil by distillation at 10^{-3} Torr and 80° C.
They were collected on a cold finger inserted in the sample flask.
Recoveries are shown in Table 2-3. Similar investigations of interest
are those of Merritt et al. (16), Angelini et al. (17), and Libbey et
al. (18).

To avoid accumulation of aqueous distillates that would require
extraction with organic solvents, Honkanen and Karvonen (19) used
carbon dioxide distillation (instead of steam) to remove flavor components from fats and oils (see Fig. 2-5). Since ether was used to
extract the distillate, only compounds boiling higher than ether were
isolated. Other extractants could be used, however, to widen the
applicability of this method.

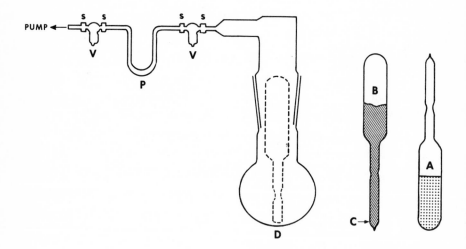

FIG. 2-4 Apparatus for recovery of C_2 through C_{10} hydrocarbons from corn oil. Reprinted from Ref. (14), p. 647,
by courtesy of the American Chemical Society.

TABLE 2-3

Recovery of Alkanes and Alkenes from Corn Oil[a]
(1 mg of hydrocarbon in 10 g of corn oil)

Compound	% Recovery	Compound	% Recovery
n-Nonane	90	n-C_{14} alkene	99
n-Nonene	85	n-Pentadecane	98
n-Decane	90	n-Pentadecene	99
n-Decene	86	n-Hexadecane	99
n-Undecane	96	n-Hexadecene	98
n-Undecene	95	n-Heptadecane	91
n-Dodecane	98	n-Heptadecene	94
n-Dodecene	98	n-Octadecane	81
n-Dodecyne	98	n-Octadecene	69
n-Tridecane	98	n-Nonadecane	60
n-Tridecene	98	n-C_{20} alkane	45
n-C_{14} alkane	99	n-C_{22} alkane	12

[a]Reprinted from Ref. (14), p. 646, by courtesy of the American Chemical Society.

Vacuum fractional distillation has been very useful in aroma research for carrying out crude fractionations. However, if a suitably stable essential oil is at hand, it can be a very powerful method for obtaining fractions of very small boiling ranges, and in some cases,

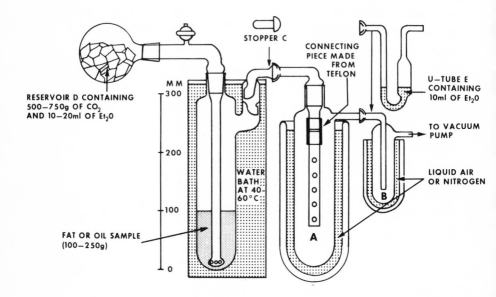

FIG. 2-5 Carbon dioxide distillation apparatus. Reprinted from
Ref. (19), p. 2626, by courtesy of the Chemical
Societies of Denmark, Finland, Norway, and Sweden.

for obtaining pure material. With the development of the Teflon spinning band (20), vacuum fractional distillation can be used for preparation of pure material, provided the material survives prolonged periods at elevated temperatures (21).

The two important parameters in distillation are temperature and
pressure [see Ref. (22) for detailed discussions]. Temperature
controls are usually adequate, but pressure controls are still very
crude. Therefore, we will briefly discuss a pressure control system
which is proving to be helpful in achieving good fractionations.

It must be pointed out that boil-up rate must be lessened as the
pressure is diminished. As a rough estimate, the vapor density at
7.6 Torr is 1/100 that at 760 Torr. Therefore, the molecules must

be moving 100 times faster up the column to yield the same amount of condensate obtained at 760 Torr over the same period of time. At pressures below 5 Torr pressure fluctuations become so serious that equilibrium conditions are not maintained. Moreover, the vapor is traveling so fast up the column that there is very little redistribution between the vapor and liquid phases. The Teflon band spinning at high speed forces the liquid to contact the vapor more than did previously available columns, and this contact is achieved with a relatively small pressure drop.

Figure 2-6 shows a diagram of a mercury manostat with enhanced sensitivity (23). The mercury, at an angle, travels further per pressure change than it would if it traveled vertically. When stopcock A is closed, the pressure for the distillation is set. Pressure changes

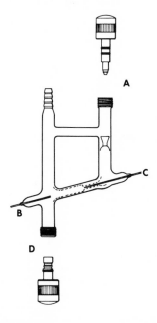

FIG. 2-6 Mercury manostat. A, High vacuum stopcock. B and C, Electrodes connected to relay. D, Screw level adjuster.

are sensed by the two electrodes, B and C, which are connected to a
relay. Minor adjustments can be made by raising or lowering the
mercury level with the screw level adjuster, D, and by tilting the manostat. The relay controls a solenoid valve (see Fig. 2-7). With this
manostat and solenoid valve system, the pressure fluctuation can be
held to less than ±0.1 Torr.

A compact solenoid valving system, Fig. 2-7, can be made from
metal parts, with connections to a glass manifold and to a vacuum
pump via ball socket joints, A and E. For rapid pumping to low pressures, the large bellows valve, B, can be opened. For pressure
control, the bellows valve is closed and the needle valve, C, is
opened so that the solenoid valve, D, controlled by the manostat and
relay, opens and closes the pump vacuum to the manifold. The
needle valve is necessary to minimize the fluctuations caused by the

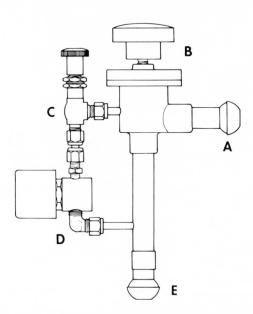

FIG. 2-7 Solenoid Valve System, A, Ball Socket Joint to Manifold.
          B, Bellows Valve. C, Needle Valve. D, Solenoid
          Valve. E, Ball Socket Joint to Pump.

solenoid valve opening and closing. The needle valve is opened wide
for evacuating the manifold and receiving flask after each fraction
change but is almost closed during the distillation. If this system is
properly manipulated, no perceptible changes are observed in the
boil-up rate or the boiling point when the receiving flask is changed or
during a prolonged distillation over several days. Only with such conditions can good vacuum fractional distillation fractionation be
achieved.

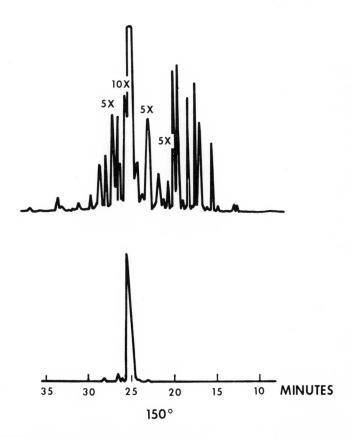

FIG. 2-8 Gas chromatograms of sesquiterpenes. Upper, mixture
from oleoresin of Pinus armandi. Lower, α-Murolene
fraction (24).

An example of what can be achieved is illustrated in Fig. 2-8. The
upper chromatogram shows a mixture of sesquiterpenes obtained from
a crude distillation of the volatiles from the oleoresin from Pinus
armandi (24). The lower chromatogram shows the high purity of the
α-murolene fraction.

The conditions for this distillation with the Teflon spinning band
column were as follows: 8 mm X 1 m band, spinning at 7200 rpm;
head vapor temperature, 95°C; pressure, 3 Torr; boil-up, 20 drops/
min; reflux ratio, 50:1. The α-murolene fraction (9 g) was obtained
from a pot charge of 60 g. The following fraction amounted to 4.5 g
and was almost as pure based on GC analysis.

3. Other Distillation Methods

The isolation of flavor concentrates from meat, bread, or other
products not easily comminuted, pressed, or otherwise converted
into essentially liquid states, is particularly challenging. Entrainment of volatiles with an inert gas in either an open or a closed system
is useful in some cases [butter, margarine, cheese (25), potato
chips (26)]. The compounds may be collected and accumulated on a
precolumn of an appropriate nature and held at suitable temperature
(27). Care must be taken to avoid accumulation (particularly in an
"open" system) of gas impurities that swamp out the naturally
occurring food volatiles.

Volatile components of cooked meat flavors (28) have been isolated
by refluxing beef (29) or chicken (30) in water. This not only
develops the desired aroma but also achieves partial solution and
suspension of the flavor materials in aqueous solutions or broths that
may be subsequently concentrated by distillation or other means.
Cooked chicken pieces have been "powdered" by grinding in liquid
nitrogen; the powder is vacuum-distilled and then mixed with water
and distilled again to achieve as complete removal of volatiles as
possible (30). Good judgment and care must be exercised in such

procedures to prevent formation of artifacts over and above formation of components of the cooked flavors.

Lyophilization or freeze drying is quite often useful for removal of volatiles from solid or semisolid materials that are not conveniently vacuum-distilled in their natural states at room temperature. In the apparatus shown in Fig. 2-9 frozen aqueous slurries of fish protein concentrate and of crushed banana pulp have yielded satisfactory quantities of aroma concentrates for detailed chemical investigation (31, 32). In such cases, as well as those in which food products are vacuum-distilled without prior addition of water, complete odor removal is never achieved. The nonvolatile residues at first appear to

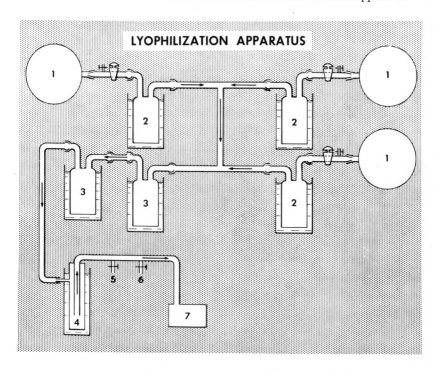

FIG. 2-9 Apparatus for lyophilization of aqueous food slurries (31).

be odorless, but as they pick up moisture from the air or if water is added to them, the original odor — or a close facsimile of it — appears.

Preliminary experiments by Weurman (33) illustrate roughly the degree to which three commonly occurring volatiles (iso-amyl acetate, iso-amyl alcohol, and diethylketone) are retained during lyophilization of mixtures of these compounds in "model" food matrices such as starch, pectin, protein, lipid, etc. The results are summarized in Table 2-4.

This retention of aroma compounds is, of course, the reason why freeze-dried, spray-dried, and other types of dried foods have flavor. In fact, the quality of such dried foods depends in large measure on the degree to which flavor volatiles have been retained in spite of the vigorous distillation procedures employed. Since the flavor research worker's purpose is to overcome this retention, it is instructive to know the principles involved. The work by Rey (34) and Menting and Hoogstad (35) and a discussion by Thijssen and Rulkens (36) of retention of aromas in drying food liquids are particularly interesting. In general, aroma retention depends strongly on the concentration of dissolved solids in the liquid to be dried. This is illustrated in Fig. 2-10, which shows that excellent retention of even highly volatile substances such as ethylene is achieved during spray drying of coffee extract which had up to 50% of initial dry matter content. Retention was attributed to formation of a dry layer permeable to water but impermeable to the volatiles. Adsorption may also play a role. No wonder it is difficult to remove aroma constituents from some food products!

In summary, distillation in its various forms (see Table 2-2) is a powerful means for separating volatiles from the mass of nonvolatile food constituents. In most cases the volatiles are collected in

TABLE 2-4

Effects of Nonvolatile Food Components on Recovery (%)
of Volatile Constituents[a]

	Diethyl-ketone, 7 ppm		Iso-amyl Acetate, 4 ppm		Iso-amyl Alcohol, 80 ppm	
	Dist[b]	Res[b]	Dist	Res	Dist	Res
In H_2O	104	—	87	—	97	—
In H_2O + 1% starch	94	5	83	9	88	4
In H_2O + 1% pectin	98	5	86	11	99	4
In H_2O + 1% cellulose	86	—	89	—	87	—
In H_2O + 1% carbon	45	—	34	—	38	—
In H_2O + 1% lipid	85	—	87	—	87	—
In H_2O + 1% protein	82	6	48	36	75	11

[a]From Ref. (33), p. 11.
[b]Dist = distillate, Res = residue.

dilute aqueous solutions. They must, therefore, be further con-centrated and isolated free of water before they can be separated and individually characterized as aroma constituents.

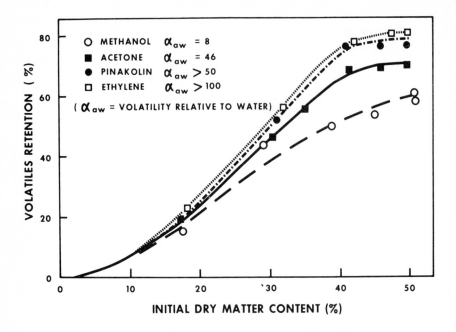

FIG. 2-10   Retention of volatiles after spray drying of coffee
             extract as a function of initial dry matter.
             Reprinted from Ref. (36), p. 51, by courtesy of
             the Netherlands Royal Institute of Engineers.

A word of warning must be included at this point: If the distillates
obtained by the methods discussed above do not exhibit the aroma
under investigation, one had best stop and consider carefully whether
the reason for the work has somehow been lost or destroyed!
Sensory evaluations that will validly assess aroma quality and intensity
require careful attention to sample preparation.   Dilution of distillates
to original concentration and comparison with the original product can
indicate the degree of success achieved in isolation of aroma
constituents.   However, one cannot expect to assess total flavor isolation since nonvolatile taste components are not present.   Reconstitu-
tion of the original product by recombination of volatile and nonvolatile
fractions can provide important information about the relative
importance of taste and odor constituents.   Without sensory evalua-

tions chemists have no guideposts and will almost certainly lose their way among the byways of flavor research.

B. Methods for Concentration of Flavor Distillates

1. Freeze Concentration

Freeze concentration is an attractive method for achieving the removal of water from dilute aqueous aroma distillates. Not only does it provide mild conditions and thus avoid artifact formation and destruction of heat-labile materials, but it is also rather simply carried out. Batches of solution may be placed in beakers, covered with foil or film, and stirred continuously below freezing temperature. Temperature adjustment and rate of stirring are critical because too-rapid freezing causes inclusion of organic constituents within the ice phase. In turn, too-rapid stirring will cause channeling in the ice due to air bubbles and possibly aroma components being trapped in the solid layer. Having once adjusted temperature and rate of stirring, the major possible cause of volatile loss is evaporation. This can be prevented, however, by covering the solution.

Kepner et al. (37) have achieved from fivefold to better than 40-fold concentration increases without appreciable losses of volatiles. They employed the beaker technique of Shapiro (38) and took 2.5-liter batches that required a 5- to 6-hr freezing period. The best concentration factors were obtained by combining successive concentrates and at intervals removing the ice phase. As shown in Table 2-5, known aqueous solutions of the composition shown were concentrated up to 6-, 15-, and 20-fold with recoveries over 90%. The amount of ethanol in the original solution often limits the degree of possible concentration.

Freeze concentration for research purposes and for possible commercial applications has been discussed by Weurman (39) and Muller (40), respectively. Applications of the method to aroma research

TABLE 2-5

Recoveries of Volatiles from Freeze Concentration of Dilute Aqueous Solutions[a]

Compound	Initial concentration, μl/liter	15-Fold concentrate, %	Initial concentration, μl/liter	Sixfold concentrate, %
I				
Ethyl formate	6	98 ± 8	20	94 ± 6
Ethyl butyrate	6	92 ± 5	20	93 ± 8
Ethyl caproate	15	96 ± 7	50	100 ± 2
3-Pentanone	12	95 ± 2	40	90 ± 3
4-Heptanone	15	96 ± 4	50	94 ± 5
3-Heptanone	15	96 ± 5	50	95 ± 1
3-Octanone	24	95 ± 6	80	100 ± 1
Ethanol	360	97 ± 5	1000	90 ± 4
n-Butanol	360	96 ± 4	–	–
II		20-Fold concentrate, %		
Ethyl caproate	18	91 ± 7		
n-Propanol	240	92 ± 9		
Isopentanol	360	92 ± 9		
n-Butanol	8	93 ± 9		
n-Hexanal	12	90 ± 8		

[a]From Ref. (37), pp. 1123, 1125.

have been surprisingly few in view of its potential usefulness. Concentrates of cocoa aroma (41), bread pre-ferment solutions (42), fruit juices (43, 44), and molasses volatiles (45) have been thus obtained. In general, the aqueous concentrates resulting from freeze concentration are excellent candidates for solvent extraction and further isolation of the organic constituents.

2. Zone Melting

Concentration of volatiles by zone melting has been reported in very few instances, though considerable conversation about the method occurs whenever aroma chemists gather. Like freeze concentration it minimizes loss of volatiles and thermal degradation. It too results in concentration of organic materials because their distribution coefficients favor the liquid rather than the frozen phase. Pfann's book (46) contains extensive treatment of the principles and theory of zone melting. In an apparatus described by Huckle (47) a 3000-fold aroma concentrate of raspberry juice was isolated. Juice (9 liters) from 50 lb of berries was extracted with 2 liters of benzene. This volume of benzene extract was concentrated (in 100-ml batches) by three-stage zone melting to a final volume of 60 μl. This concentrate was described as containing essentially no impurities (because zone melting requires use of "ultrapure" solvents) and exhibiting an unusually "fresh" aroma. Concentrates of vegetable and fruit flavors and of flower perfumes are reported to have been obtained in this manner. Cyclohexane, as well as benzene, is suitable for use in zone melting (48).

Though zone melting was very useful in the above cases, it appears not to be widely applicable due to the complexity of the required apparatus, the restriction to small sample size, the need to use solvents that have already been purified by zone melting, and the time required to accomplish the overall concentration.

3. Adsorption

Another attractive means for isolating aroma volatiles free of large volumes of water or organic solvent is adsorption. Charcoal is most often used for this purpose because it is not deactivated by water and it has a large adsorptive capacity. Aroma volatiles from aqueous pea (49), pear (50), onion (51), and apple (52) essences have been successfully concentrated on charcoal. Different types of charcoal reach adsorption equilibrium after different periods of use and differ somewhat in their selectivity and capacity. Thus, the choice of type to be used is important. In general, adsorption of organic substances on carbon from solvents decreases in the following order: water, ethanol, esters, acetone, chloroform (39).

Possible artifact formation and rearrangement or interaction between materials held on charcoal have been mentioned by various investigators (50, 51, 53). Although, in these cases, no evidence was found that such reactions took place, their occurrence is likely.

Excess water may be removed from charcoal that contains adsorbed materials either by heating or lyophilization. The latter procedure is preferred because of its mild conditions. No loss of pear volatiles was observed (50) when this treatment was used.

Alternatively, the wet charcoal may be immediately eluted with organic solvent. Diethyl ether has been used most commonly for elution of flavor components. Direct ether elution, without drying the adsorbent and without continuous extraction in a Soxhlet, was successful in desorbing apple volatiles (52). Carbon disulfide has also been used satisfactorily as an eluant (54, 55).

For researchers armed with the knowledge that types of charcoal differ in their selectivity and that some components may be irreversibly held, charcoal adsorption can be a powerful tool for the isolation of aroma concentrates.

4. Adsorption Chromatography

When one's starting material is an essential oil and thus is already an "aroma concentrate," further separation and isolation of more purified fractions may be accomplished by adsorption chromatography. The separation of hydrocarbons from oxygenated components is an example of the usefulness of silica gel chromatography for this purpose. Murray and Stanley (56) have employed silica gel columns with ascending solvent flow to separate ester, ketone, and alcohol components of pea and banana aroma concentrates (57, 58). The resulting relatively "simple mixtures" were subjected to gas chromatography-mass spectrometry (GC-MS) analysis. Modification of this technique for descending elution significantly improved the separation of the organic classes (59). In addition, elution with odorless Freon 11 (bp 23° C) facilitated aroma evaluation of the eluted groups of compounds.

A related but somewhat different procedure is that of "dry column chromatography" (60). The method is applicable to 1- to 50-g quantities. The separations achieved are comparable to those obtainable by thin layer chromatography (TLC) and can be accomplished rapidly. An empty column is filled with "suitably deactivated adsorbent" and the sample to be separated is complete when the developing solvent makes its way to the bottom of the dry column (15-30 min). Separated fractions are then removed from the column. Optimum conditions for a given separation are developed on TLC plates or chromatostrips. If the same solvent and same adsorbent are used in the dry column, a similar degree of separation is achieved. Deactivation of adsorbents (alumina or silica gel) to a known and reproducible Brockmann activity grade (II-III) is the key to success in the dry column method.

5. Extraction

Although extraction as a method for isolation of flavor concentrates has been mentioned several times, the subject has been purposely postponed until now. The reason is that all practical alternative concentration procedures should first be considered so that maximum effectiveness may be gained from solvent extraction. In a survey of the literature, Weurman (39) found that in actual practice, extraction has been used by most flavor investigators as the first treatment applied after initial distillative separation of volatiles from nonvolatiles. He points out the questionable logic of this approach. Direct extraction (whether batch or continuous) of large volumes of dilute solution can only result in large volumes of dilute extractant from which the desired aroma components must still somehow be isolated and concentrated. During this subsequent concentration, solvent impurities are accumulated along with the flavor constituents and thus they complicate the problem unnecessarily. It is therefore very important to achieve as much concentration as possible before extraction procedures are applied.

Whenever practical, extraction solvents should be chosen on the basis of their selectivity for the compounds of interest. Unfortunately, in flavor research one rarely knows ahead of time the polarity or functionality of the important constituents. However, one can often make a good guess as to what one does not wish to isolate. Ethanol, for example, is often a major constituent that can profitably be exempted from the extracted isolate. Saturation of aqueous solutions with salts often contributes to the completeness of extraction though it may diminish selectivity of the extraction solvent. Commonly used salts are sodium sulfate and sodium chloride.

A comprehensive list of organic solvents that may be useful for extraction of aroma volatiles is given by Weurman (39). With liquids such as the pentanes and fluorocarbons, ethanol is largely excluded from the solvent extracts. Since ethanol and other alkanols are often the major volatile organic food constituents but make little or no flavor contribution, methods for avoiding their extraction along with the active flavoring ingredients are highly desirable.

The suitability of trichlorofluoromethane (Freon 11) as a solvent
that excludes alcohols but extracts other organic materials has been
investigated by Hardy (61). Because Freon 11 (bp 23° C) is nontoxic,
nonflammable, nonexplosive, and essentially odorless, its use is
advantageous in flavor research (60). Continuous extraction at 10° C
with Freon 11 was carried out on various aqueous solutions of known
composition (61). Above C_4 extraction efficiency improved greatly.
In the ethanol solutions experimental difficulties prevented measurement of compounds below C_5. Between 70 and 100% of the alkanols,
2-alkanones, and ethyl alkanoates between 5 and 12 carbon number
were recovered. Due to the insolubility of water in Freon 11, no
drying of the extracts was required before concentration. It thus
appears that this solvent may have valuable applications to flavor
investigations. The commercial use of chlorofluorocarbon solvents
for extraction of flavors (62, 63) and fragrances (64) has been discussed
and is considered to yield water-free isolates of high flavor quality.

Another seldom-used solvent for extraction of flavor volatiles from
dilute aqueous solution is paraffin oil. Its advantages are inertness,
immiscibility with water, low cost, and nonvolatility. Nelson and Hoff
(65) developed a paraffin oil extraction procedure to determine the
concentration of selected volatile tomato constituents (66). The per
cent recovery of selected components from aqueous solution and from
the paraffin oil by gas stripping is summarized in Table 2-6.

A very interesting comparison of the effectiveness of various
extractants for isolation of volatiles from apple essence has been made
by Schultz et al. (52). The extractants were isopentane, diethyl ether,
charcoal (with elution by ether), liquid carbon dioxide, and 1, 2-dichloro-
1, 1, 2, 2-tetrafluoroethane (bp 4° C). Quantitative estimation of major
components in the extracts and sensory odor evaluations of the whole
extracts were made. Identification of components was accomplished
by GC-MS analysis. The authors state: "isopentane and the

TABLE 2-6[a]

A. Per cent Recovery of Certain Compounds from
Aqueous Solution (300 ml)

Compound	Amount present, ml	% Recovery, average
Acetone	6.71×10^{-4}	34
Methanol	1.7×10^{-3}	4.1
Ethyl acetate	3.75×10^{-4}	73
Isovaleraldehyde	2.8×10^{-4}	56
2-Pentanone	5.19×10^{-4}	57

B. Per cent Recovery from Paraffin Oil after
Gas Stripping at 40°C

Compound	% Recovery, uncorrected
Acetaldehyde	91.5
Methyl sulfide	100
Acetone	100
Methanol	69
Ethanol	100
Isovaleraldehyde	97
2-Pentanone	94
Hexanal	67

[a]Reprinted from Ref. (65), p. 1396, by courtesy of the
Institute of Food Technologists.

fluorocarbon were similar in action. Compared with ether they gave
lower recoveries, especially for the low-molecular-weight alcohols,
but they gave extracts with higher concentrations of esters and aldehydes .... Liquid carbon dioxide gave an extract similar to that given
by ether. Adsorption on charcoal gave generally good recoveries and
was outstanding for alcohols" (see Table 2-7). Panel evaluations
showed that the isopentane and fluorocarbon extracts possessed aromas
most like that of freshly pressed apple juice. However, all the extracts had a strong apple aroma (52).

A somewhat less intensive comparison of various methods for
preparation of Bartlett pear aroma concentrates is shown in Fig. 2-11
(50). Examples of such comparisons with attention given to aroma
quality are extremely rare in the literature. They are also rare in
practice, since one's tendency is to take shortcuts at every opportunity.
In the case of flavor research, however, shortcuts taken without
adequate prior study of their sensory and analytical consequences
often lead to dead ends.

Many types of apparatus are available in which liquid-liquid extractions may be accomplished. One was shown in Fig. 2-3. It illustrates
simultaneous steam distillation and continuous solvent extraction of
the distillate. The relatively small volumes of both solvent and distillate are noteworthy.

When many gallons of an aqueous essence must be extracted, a
liquid-liquid countercurrent extractor should be used (50, 67).
Figure 2-12 shows a typical vertical extractor in which the solvent and
essence are intimately mixed by a stirrer. The phases which are not
stirred separate at the lower and upper zones and are taken away from
the extractor as indicated.

If a solvent, such as chloroform or a fluorocarbon, is used in which
emulsions are formed and are not easily broken, a vertical extractor

TABLE 2-7

Extraction of Volatiles from Aqueous Apple Essence[a]

Extractant	Isopentane	Ether	Charcoal[b]	CO_2	$CClF_2-CClF_2$
Yield, g from 3 kg aqueous essence	2.53	6.05	8.41	2.96[c]	1.36
Residual solvent, %	55	40	44	< 1	~ 10
Calculated net yield, g/kg aqueous essence	0.38	1.21	1.58	0.98	0.41

[a] Reprinted from Ref. (52), p. 280, by courtesy of the Institute of Food Technologists.

[b] With elution by ether.

[c] Calculated from actual yield of 9.55 g from 9.67 kg of aqueous essence.

Chromatogram of essence obtained by direct fruit extraction.

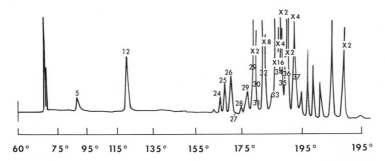

Chromatogram of essence obtained by extraction of a laboratory steam distillate.

60°   75°   95°   115°   135°   155°   175°   195°   195°

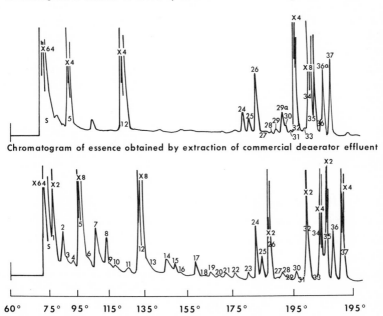

Chromatogram of essence obtained by extraction of commercial deaerator effluent

Chromatogram of essence obtained by charcoal adsorption and Soxhlet extraction

60°   75°   95°   115°   135°   155°   175°   195°   195°

FIG. 2-11   Chromatograms of pear essences obtained by various
isolation procedures. Reprinted from Ref. (50), p. 4,
by courtesy of the Institute of Food Technologists.

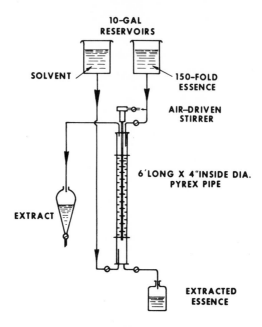

FIG. 2-12   Countercurrent liquid-liquid extractor. Reprinted
              from Ref. (67), p. 31, by courtesy of the American
              Chemical Society.

is not feasible. For such extractions a horizontal extractor, which
utilizes a large surface area and perforated stainless steel tubes that
slowly carry one liquid through another, has been designed (see Fig.
2-13). The two liquids are forced to move countercurrently. With
this type of extractor, it is possible to extract juices without the formation of emulsions, which are extremely difficult to break if carbohydrates and proteins are present in the solution (68).

Liquid carbon dioxide does require special high pressure equipment
(69), but the extraction with liquid carbon dioxide yields material
similar to that obtained with ether (52). The elimination of the solvent
in this case is accomplished under very gentle conditions, i.e.,
release of pressure at room temperature or lower. Moreover, the
extracted material can be consumed without any worry as to the

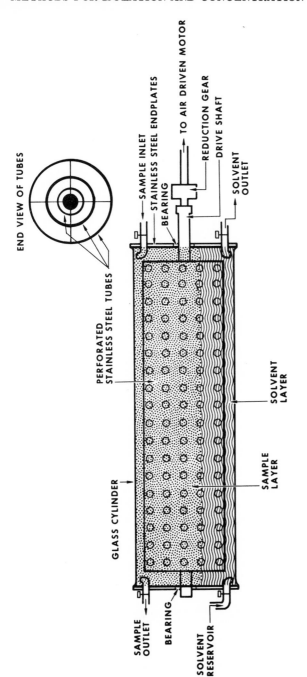

FIG. 2-13   Horizontal countercurrent liquid–liquid extractor. Reprinted from Ref. (68), p. 1102, by courtesy of the American Chemical Society.

amount of the residual extractant. This type of extraction should have some interesting applications in the food industry.

III. SUMMARY

Consideration of procedures for isolation of flavor concentrates has led to a wide-ranging review of separation and purification methods. No attempt was made to encompass all methods ever used for this purpose. For example, no mention was made of derivative formation as a means to isolate volatiles — though much work has been done in this field. The goal was, instead, to focus on techniques that can allow isolation of volatiles in their natural state so that their aroma and flavor qualities as well as their chemical identification may be determined. The strategy and tactics required to achieve this goal are dependent on the nature of the food to be investigated. Several different separation methods will, no doubt, be necessary, and artifact formation must be minimized if it cannot be prevented. Finally, successful isolation of a concentrate can only be claimed if by sensory evaluation it is demonstrated to exhibit appropriate flavor quality.

REFERENCES

1.   R. L. Hall, Food Technol., 22, 1388 (1968)

2.   L. H. Walker, in Fruit and Vegetable Processing Technology (D. K. Roessler and M. A. Joslyn, eds.), Avi, Westport, Conn., 1961, p. 358.

3.   N. F. Roger and V. A. Turkot, Food Technol., 19, 69 (1965).

4.   Y. Mälkki and J. Veldstra, Food Technol., 21, 15 (1967).

5.   J. L. Bomben, J. A. Kitson, and A. I. Morgan, Jr., Food Technol., 20, 1219 (1966)

6.   G. B. Nickerson and S. T. Likens, J. Chromatog., 21, 1 (1966).

7.   D. A. Forss, V. M. Jacobsen, and E. H. Ramshaw, J. Agr. Food Chem., 15, 1104 (1967).

8.   S. S. Chang, J. Am. Oil Chemist's Soc., 38, 669 (1961).

9.   E. A. Day and D. A. Lilliard, J. Dairy Sci., 43, 585 (1960).

10.   C. H. Lea and P. A. T. Swaboda, J. Sci. Food Agr., 13, 148 (1962).

11.   S. Patton and B. W. Tharp, J. Dairy Sci., 42, 49 (1959).

12.   W. W. Nawar and I. S. Fagerson, Food Technol., 16, 107 (1962).

13.   D. A. Forss and G. L. Holloway, J. Am. Oil Chemist's Soc., 44, 572 (1967)

14.   W. W. Nawar, J. R. Champagne, M. F. Dubravcic, and P. R. LeTellier, J. Agr. Food Chem., 17, 645 (1969).

15.   J. de Bruyn and J. C. M. Schogt, J. Am. Oil Chemist's Soc., 38, 40 (1961).

16.   C. Merritt, Jr., P. Angelini, M. L. Bazinet, and D. J. McAdoo, Advan. Chem. Ser., 56, 225 (1966).

17.   P. Angelini, D. A. Forss, M. L. Bazinet, and C. Merritt, Jr., J. Am. Oil Chemist's Soc., 44, 26 (1967).

18.   L. M. Libbey, D. D. Bills, and E. A. Day, J. Food Sci., 28, 329 (1963).

19.   E. Honkanen and P. Karvonen, Acta Chem. Scand., 20, 2626 (1966).

20.   Nester/Faust Manufacturing Corp., Wilmington, Del.

21.   This and the following information about the spinning band column has been contributed by R. Teranishi.

22.   Distillation, Techniques of Organic Chemistry (A. Weissberger, ed.), Vol. IV, Wiley (Interscience), New York-London, 1951.

23.   C. W. Ferry, Ind. Eng. Chem., Anal. Ed., 10, 647 (1938).

24.   R. Teranishi, unpublished results.

25.   M. E. Morgan and E. A. Day, J. Dairy Sci., 48, 1382 (1965).

26.   W. W. Nawar and I. S. Fagerson, Anal. Chem., 32, 1534 (1960)

27. I. Hornstein and P. F. Crowe, Anal. Chem., 34, 1354 (1962).

28. I. Hornstein (Chap. 10) and E. L. Pippen (Chap. 11), in
     Chemistry and Physiology of Flavor (H. W. Schultz, E. A. Day,
     and L. M. Libbey, eds.), Avi, Westport, Conn., 1967.

29. C. H. T. Tonsbeek, A. J. Plancken, and T. van der Weerdhof,
     J. Agr. Food Chem., 16, 1016 (1968).

30. C. H. Lea, P. A. T. Swoboda and A. Hobson-Frohock, J. Sci.
     Food Agr., 18, 245 (1967).

31. E. L. Wick, E. Underriner, and E. Paneras, J. Food Sci.,
     32, 365 (1967).

32. E. L. Wick, A. I. McCarthy, M. Myers, E. Murray,
     H. Nursten, and P. Issenberg, Advan. Chem. Ser., 56, 241
     (1966).

33. C. Weurman, seminar at U. S. Army Natick Labs., 1967.

34. L. R. Rey, Lebensm. Gefriertrocknung, Gefriertrocknungstagung,
     Koln, 3 (1962).

35. L. C. Menting and B. Hoogstad, J. Food Sci., 32, 87 (1967).

36. H. A. C. Thijssen and W. H. Rulkens, De Ingenier, 80, 45
     (1968).

37. R. E. Kepner, S. van Straten, and C. Weurman, J. Agr. Food
     Chem., 17, 1123 (1969).

38. J. Shapiro, Anal. Chem., 39, 280 (1967).

39. C. Weurman, J. Agr. Food Chem., 17, 370 (1969).

40. J. G. Muller, Food Technol., 21, 49 (1967).

41. M. van Praag, H. S. Stein, and M. S. Tibbetts, J. Agr. Food
     Chem., 16, 1005 (1968).

42. W. S. Hale and E. W. Cole, Cereal Chem., 40, 287 (1963).

43. D. S. Bidmead, in Recent Advances in Food Science (J. M. Leitch
     and D. N. Rhodes, eds.), Vol. 3, Butterworth, London, 1963, p. 158.

44. G. Senn, Z. Lebensm. Untersuch. Forsch., 120, 455 (1963).

45. C. Weurman, Proc. 2nd Intern. Congr. Food Sci. Technol.,
     Warsaw 1966, p. 289.

46. W. G. Pfann, Zone Melting, Wiley, New York, 1958.

47. M. T. Huckle, Chem. Ind. , 83, 1490 (1966).

48. R. P. W. Scott, Chem Ind. , 86, 797 (1969).

49. J. W. Ralls, W. H. McFadden, R. M. Siefert, D. R. Black, and P. W. Kilpatrick, J. Food Sci. , 30, 228 (1965).

50. D. E. Heinz, M. R. Sevenants, and W. G. Jennings, J. Food Sci. , 31, 63 (1966).

51. J. F. Carson and F. F. Wong, J. Agr. Food Chem. , 9, 140 (1961).

52. T. H. Schultz, R. A. Flath, D. R. Black, D. G. Guadagni, W. G. Schultz, and R. Teranishi, J. Food Sci. , 32, 279 (1967).

53. N. Paillard, Fruits, 20, 189 (1965).

54. C. S. Tang and W. G. Jennings, J. Agr. Food Chem. , 15, 24 (1967).

55. W. G. Jennings and H. E. Nursten, Anal. Chem. , 39, 521 (1967).

56. K. E. Murray and G. Stanley, J. Chromatog. , 34, 174 (1968).

57. K. E. Murray, J. Shipton, F. B. Whitfield, B. H. Kennett, and G. Stanley, J. Food Sci. , 33, 290 (1968).

58. K. E. Murray, J. K. Palmer, F. B. Whitfield, B. H. Kennett, and G. Stanley, J. Food Sci. , 33, 632 (1968).

59. J. K. Palmer, personal communications.

60. B. Loev and M. M. Goodman, Chem. Ind. , 1967, 2026.

61. P. J. Hardy, J. Agr. Food Chem. , 17, 656 (1969).

62. W. L. Stanley, J. E. Brekke, and R. Teranishi, U.S. Pat. 3,113,031 (1963).

63. J. Schiede and K. Bauer (to Haarmann and Reimer GmbH), British Pat. 1,069,810 (1967).

64. J. Safrin and E. J. Strobl (to Albert Verley & Co.,), U. S. Pat. 3,150,050 (1964).

65. P. E. Nelson and J. E. Hoff, Food Technol. , 22, 1395 (1968).

66.   P. E. Nelson and J. E. Hoff, J. Food Sci. , 34, 53 (1969).

67.   R. A. Flath, D. R. Black, D. G. Guadagni, W. H. McFadden, and T. H. Schultz, J. Agr. Food Chem. , 15, 29 (1967).

68.   K. L. Stevens, R. A. Flath, A. Lee, and D. J. Stem, J. Agr. Food Chem. , 17, 1102 (1969).

69.   G. Schultz, paper presented at 26th Annual Meeting, Institute of Food Technologists, Portland, Ore. , May, 1966.

Chapter 3

GAS CHROMATOGRAPHY SEPARATIONS

INTRODUCTION

Aroma chemistry analyses have special requirements. Constituents must be separated, detected, and identified from complex mixtures which are usually available only in small amounts. The constituents must be characterized with regard not only to detailed chemical structure and physical properties but also to sensory properties. This extension demands that fractions be more than spectrometrically pure, which is difficult enough to achieve — they must also be sensorially pure. Most spectrometers cannot detect less than 1% of an impurity, but a fraction of a per cent of β-ionone would have an appreciable sensory effect in an n-amyl acetate sample [see thresholds in Table 3-1 (1)]. Moreover, thresholds vary over many orders of magnitude, and our analytical methods must cover this wide range. Also, the analyses must be effective for a wide variety of compounds: free

TABLE 3-1

Odor Thresholds[a]

Odorant	In water, μg/liter
Formic acid	450,000
Ethanol	100,000
Propanol	9,000
Butyric acid	240
Nootkatone	170
n-Amyl acetate	5
Dimethyl sulfide	0.3
n-Decanal	0.1
Methyl mercaptan	0.02
β-Ionone	0.007
2-Methoxy-3-isobutylpyrazine	0.002

[a] D. G. Guadagni, private communication, 1969.

amines and acids, nitrogen and sulfur compounds, very stable and
very labile compounds, permanent gases, and compounds with little
vapor pressure, etc. We cannot concentrate our attention on any
single group of compounds.

Haagen-Smit (2) has pointed out that the basic steps in analysis are
separation and identification. This concept is given in outline form in
Table 3-2. The most frequently used methods of separation in aroma
research are distillation and gas chromatography (GC). Distillation is
primarily a separation method (see Chapter 2), but boiling points can
be used as indications of what range of compounds may be present.
GC is also primarily a separation method. Although retention data

TABLE 3-2

Analytical Procedure

I. Separation and isolation

   A. Distillation

   B. Crystallization

   C. Chromatography

     1. Liquid

     2. Gas

II. Detection

   A. Physical

   B. Chemical

   C. Biological

III. Identification and characterization

   A. Chemical Structure

     1. Chromatographic retention

     2. Spectrometric

       a. Ultraviolet

       b. Infrared

       c. Raman

       d. Mass

       e. Nuclear magnetic resonance

     3. Chemical degradation and synthesis

   B. Biological activity

---

can be used for identification, such data should be used only to indicate the range of compounds which may be present. GC has developed rapidly and enjoys widespread use because of two excellent methods of detection: thermal conductivity (physical) and flame ionization (chemical). In aroma chemistry, obviously, biological detection is achieved by sniffing the effluents from a GC instrument equipped with a nondestructive detector. For a good degree of certainty of identification, spectrometric data must be obtained. For the best degree of identification, chemical degradation to known compounds and total and unequivocal synthesis must be achieved. In flavor research, biological activity of an isolated fraction must be established by responses of experts or of panels of trained individuals, or both.

The concept of the degree of certainty of identification and characterization is often neglected. Ralls (3) has related this concept to

methods and amounts necessary, as shown in Fig. 3-1. There are no
clear boundaries, and it is even debatable as to which method is the
most powerful for identification in any particular case. Also, as
methods and equipment are improved, the amounts necessary will be
reduced. Nevertheless, the general concept remains valid, and the
diagram in Fig. 3-1 gives perspective on how methods fit into the
general scheme of identification. It points out that we must always
question just how tenuous or secure our identifications are.

With small molecular weight, easily separated, well-described
compounds, it may be sufficient to use chromatographic retention
data, such as obtained from paper chromatography (PC), thin layer
chromatography (TLC), or gas chromatography (GC). Retention data
become much more valuable when used with other information, such as
mass spectra. If compositional studies previously have been made
with spectral and chemical data for identifications, then retention data
studies can provide a fair degree of certainty. However, with an unknown, undescribed compound, it must be emphasized that it is necessary to degrade it to a known compound and to synthesize it via
unequivocal reactions for a secure identification. Such work requires
much more material and effort than that required for mere retention
studies, but for a good degree of certainty these requirements must be
met.

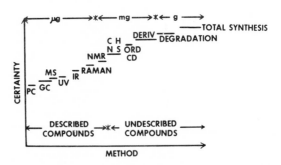

FIG. 3-1 Certainty versus method, From Ref. (3).

Certainty in sensory characterization is even more difficult to establish than structural characterization. Even if we assume that the difficulties in describing sensory responses qualitatively and quantitatively have been surmounted, we must always question the sensory purity of the sample. Although there have been remarkable advances in separation methods, we must continue to improve and extend them in order to obtain sensorially pure samples.

Thus, each time we do a separation, we must ask: (1) How complex is the mixture? (2) How much material must be separated and isolated to attain the desired degree of certainty of characterization? and (3) Has sensory purity or even spectrometric purity been achieved? This discussion of GC is not a general, comprehensive review. Instead, it expresses the viewpoint of investigators involved in aroma research who are trying to answer the above questions.

II. RESOLUTION

We will discuss some parameters which are pertinent and important in obtaining resolution specifically needed in aroma research. A few references to general discussions of GC are listed (4-10).

Chromatography is primarily a separation process which depends on the redistribution of the molecules of a mixture between two or more phases. One is usually a thin phase and the other a bulk phase brought into contact in a differential, countercurrent manner (11). Thus, gas chromatography, introduced by James and Martin in 1952 (12), is a separation method which depends on the redistribution of molecules between the gaseous phase and the liquid or solid phase. Although the ultimate achievement, the separation of enantiomers, has been reported by Gil-Av et al. (13), most GC separations fall far short of such an elegant accomplishment.

James and Martin (12) pointed out in a qualitative manner the factors important in resolution. Dal Nogare (14) summarized the importance of tightness of bands (column efficiency) and selectivity in

the expression shown in Fig. 3-2, in which the terms are succinctly
defined. The expression for resolution can be simply derived from the
commonly accepted definition of resolution, R:

$$R = \frac{x_2 - x_1}{w} \tag{1}$$

where x_2 and x_1 are the elution times of peaks 2 and 1, respectively,
and w is the average peak width of x_2 and x_1. This definition of resolution is intuitively acceptable since it tells us approximately how many
peaks of similar width can be fitted between peaks x_2 and x_1. By
using the definition of the number of theoretical plates, N:

$$N = 16 \left(\frac{x_1}{w} \right)^2 \tag{2}$$

and by substituting for w in Eq. (1) its value from Eq. (2), the
expression for resolution, Fig. 3-2, is obtained:

$$R = \frac{\sqrt{N}}{4} \left[\frac{x_2}{x_1} - 1 \right] \tag{3}$$

This expression states that GC resolution is proportional to the square
root of the number of theoretical plates (TP) and directly proportional
to selectivity, which is simply defined as the ratio of the retention
times, x_2/x_1. The concept expressed qualitatively by James and
Martin (12), and algebraically by Dal Nogare (14), has been shown
pictorially by Littlewood (5) (Fig. 3-3). Resolution can be improved
by a change in selectivity, chromatogram (b), or by keeping the same
selectivity but making a column with less band broadening, chromatogram (c).

If we have a mixture of a few compounds, changes in selectivity
can be used to improve resolution. In aroma research, we often have
mixtures containing hundreds of compounds. In such cases, different
selectivities would only change retention times such that different
mixtures would be found in any peak. The effectiveness of tightness

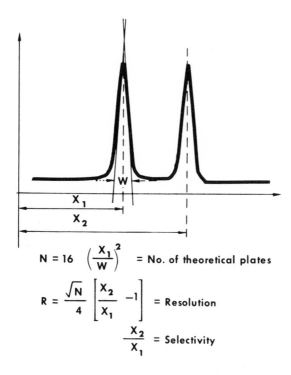

$$N = 16 \left(\frac{X_1}{W} \right)^2 = \text{No. of theoretical plates}$$

$$R = \frac{\sqrt{N}}{4} \left[\frac{X_2}{X_1} - 1 \right] = \text{Resolution}$$

$$\frac{X_2}{X_1} = \text{Selectivity}$$

FIG. 3-2   Definitions of number of theoretical plates, resolu-
            tion, and selectivity. From Ref. (14).

of bands has been shown by Buttery et al. (15) in the separation of
sesquiterpenes (Fig. 3-4). Peak A from the packed column has been
shown to contain two compounds, α-ylangene and α-copaene, peaks
41 and 42 from the capillary column. Because the selectivities of
these very closely related compounds do not change much with different
stationary liquids, it is necessary in aroma research to use columns
with large numbers of theoretical plates, i.e., columns in which
there is little band spreading.

A. Tightness of Bands

The factors contributing to band spreading are summarized in the
van Deemter-Golay equation, Eq. (4):

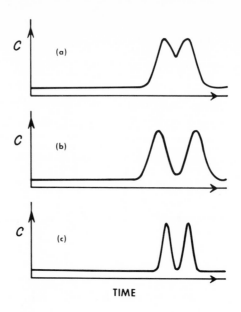

FIG. 3-3   Better resolution by different selectivity or by tighter
bands shown pictorially. From Ref. (5).

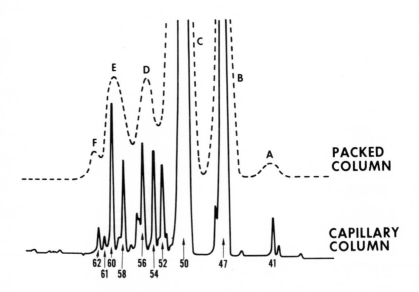

FIG. 3-4   Separation of sesquiterpenes with packed and
capillary columns. From Ref. (15).

84

$$H = Ad_{part} + \frac{B}{u} + C_g u + C_1 u \tag{4}$$

where H is the height equivalent to a theoretical plate; Ad_{part} is the band spreading from the multiple-path effect, proportional to the diameter of the particles of the solid support, d_{part}; B/u is the band spreading from molecular diffusion; $C_g u$ is the band spreading from resistance to mass transfer in the gas phase; $C_1 u$ is the band spreading from resistance to mass transfer in the liquid phase; and u is the average linear velocity of the carrier gas. In further detail, the resistance to mass transfer terms, Eq. (4), are expressed as:

$$C_g = \left(\frac{1 + 6k + 11k^2}{24 (1 + k)^2} \right) \left(\frac{r^2}{D_g} \right) \tag{5}$$

$$C_1 = \left(\frac{2k}{3 (1 + k)^2} \right) \left(\frac{d_1^2}{D_1} \right) \tag{6}$$

where k is a function of the partition coefficient, r is the inside radius of the open tubular column, d_1 is the liquid film thickness, and D_g and D_1 are the diffusivities of the solute in the gas and liquid phases, respectively.

If some of the experimental details from Eqs. (5) and (6) are incorporated in Eq. (4), it becomes:

$$H = Ad_{part} + \frac{B}{u} + C'_g \frac{r^2}{D_g} u + C'_1 \frac{d_1^2}{D_1} u \tag{7}$$

In packed columns, the band spreading from the multiple-path effect, Ad_{part}, is minimized by using the finest and most uniform solid support particles permitted by a practical pressure drop. Of course, with open tubular columns, this term is zero. The band spreading from molecular diffusion, B/u, is minimized by keeping the average linear velocity of the carrier gas more than 3 cm/sec with packed columns and more than 15 cm/sec with open tubular columns. Band spreading from resistance to mass transfer in the gas phase, third term in Eq. (7), is minimized by keeping the radius of the open

tubular column, r, as small as possible. Stainless steel tubing is
commercially available in 0.01-, 0.02-, and 0.03-in., i.d., sizes.
This third term predicts that the highest efficiency would be achieved
with the 0.01-in.-i.d. tubing, but in practice, the 0.02- and 0.03-in.-
i.d. tubings yield almost as many TP per foot as the 0.01-in.-i.d.
tubing. Therefore, because of the ease of washing and coating and of
the capacity of the larger bore tubing, the 0.02- and 0.03-in.-i.d.
open tubular columns are more practical. Because the band spreading
from resistance to mass transfer in the liquid phase, fourth term in
Eq. (7), increases as the square of the film thickness, d_1, it is
imperative to spread the liquid phase thinly and evenly. Also, in
order to obtain columns with over 100,000 TP, it is necessary to use
liquid phases of low viscosity, high diffusivity, or large D_1. This
contribution to band spreading by low diffusivity in the liquid phase is
not significant with packed columns which have only several thousand
TP. With open tubular columns, one can observe easily the change in
viscosity as the temperature of the column is raised. Near the melting point of the liquid phase, a column may have only several thousand
TP. However, at elevated temperatures, the viscosity of the liquid
phase is low, the diffusivity of the solute in it is high, and the same
column may have several hundred thousand TP.

Equation (7) shows that the band spreading will be minimal at some
average linear velocity of the carrier gas, u. It is instructive to
examine an experimental van Deemter curve, Fig. 3-5. It is apparent
that the important experimental parameter is the average linear
velocity of the carrier gas and not the flow volume usually measured.
This fact becomes very important if columns of various diameters are
used. Moreover, it is easier to obtain the average linear velocity
because all that is measured is the elution time of unabsorbed material: air peak for thermal conductivity and methane peak for flame
ionization detectors. With such data, experimental conditions for
optimum column efficiency can be easily set. The inlet pressure of

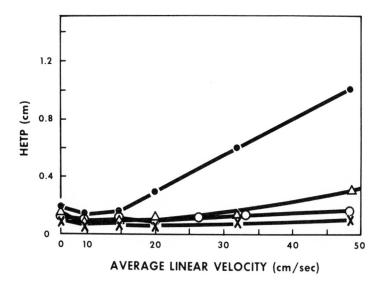

FIG. 3-5   Experimental van Deemter curve, From Ref. (33).

the carrier gas can be adjusted for the proper linear velocity. For very small samples, linear velocities higher than optimum can be used without serious band spreading. For preparative separations, however, it is imperative to set the linear velocity at the optimum conditions or serious band spreading will be observed. This effect is not usually observed with columns with low TP values.

B. Selectivity

The practical significance in aroma research of columns with large numbers of TP has been illustrated in Fig. 3-4. However, it must be kept in mind that selectivity is very important in all separations. In some cases, compounds which are very difficult to separate with open tubular columns coated with a given liquid phase can be easily separated with a packed column with a different liquid phase. Although there are many liquid phases commercially available, there has been little effort in quantitatively evaluating selectivities (5,16). Because

polarity classification is somewhat involved (16), it is not common
practice to evaluate columns in such a manner. However, since the
number of TP and selectivities do play an important part in resolution,
some evaluation of selectivities should be made.

Mon et al. (17,18) used a mixture of compounds representative of
those encountered in their work to calculate TP and selectivities. TP
are calculated with n-hexanol, n-octanal, n-amyl acetate, limonene,
and n-decane. Relative retention times (RRT) with respect to n-decane
(5) give a measure of selectivities of the liquid phases with direct
meaning to the experimenter. Table 3-3 shows a range of selectivities
illustrated by compounds of different functional groups in the test mixture with different liquid phases. These RRT values show what a
powerful means of separation the use of selectivity is and why all possible stationary phases must be available for the difficult separations
encountered in aroma research. This RRT information becomes more
valuable as more is known about the functionality of the compounds
present in mixtures.

Table 3-4 compares RRT values of the test compounds using packed
and open tubular columns. With polyethylene glycol, Carbowax 20M,
the RRT's with packed and open tubular columns are very similar.
However, with a hydrocarbon such as purified apiezon L, there is
considerable difference. The difference accounts for the difficulties in
correlation of packed column retention times with open tubular column
retention times. Table 3-5 shows the change in RRT with various
treatments to reduce adsorption on the solid support. These results
show how much selectivity of a nonpolar liquid phase can change with
the amount of adsorption. Reporting just the liquid phase does not
adequately define the selectivity.

In separations of complex mixtures, different amounts of adsorption can change the elution times enough to give confusing results.
Figure 3-6 shows separation with two "identical" open tubular columns,
both made from tubing purchased at the same time from a manufacturer,

TABLE 3-3

Relative Retention Times (Open Tubular)

	L (purified)	L	96(50)	20M	T-20	X-305
n-Decane	1.0	1.0	1.0	1.0	1.0	1.0
Limonene	1.1	1.2	1.2	3.0	2.9	4.0
n-Amyl acetate	0.38	0.73	0.53	2.2	2.0	3.0
n-Octanal	0.58	0.93	0.94	4.2	4.3	5.6
n-Hexanol	0.22	0.63	0.44	4.5	4.9	14.0

TABLE 3-4

Relative Retention Times

	L (purified)		96(50)		20M	
	Open tubular	Packed	Open tubular	Packed	Open tubular	Packed
n-Decane	1.0	1.0	1.0	1.0	1.0	1.0
Limonene	1.1	1.4	1.2	1.2	3.0	3.1
n-Amyl acetate	0.38	0.46	0.53	0.60	2.2	2.4
n-Octanal	0.58	0.89	0.94	1.0	4.2	3.8
n-Hexanol	0.22	0.50	0.44	0.52	4.5	4.4

and both cleaned and coated simultaneously under "identical" conditions by one experimenter. Yet sensorially important compounds (19), trans-2-hexenal and ethyl 2-methylbutyrate, elute in reverse order.

TABLE 3-5

Relative Retention Times with SF96 (50)

	Open tubular	Packed 1% SF96(50) 0.05% Igepal HMDS G	Packed 1% SF96(50) 0.01% Igepal HMDS G	Packed 1% SF96(50) 0.05% Igepal No HMDS	Packed 4% SF96(50) G with water	Packed 4% SF96(50) G without water
n-Decane	1.0	1.0	1.0	1.0	1.0	1.0
Limonene	1.2	1.2	1.2	1.2	1.2	1.2
n-Amyl acetate	0.53	0.55	0.56	0.56	0.60	0.77
n-Octanal	0.94	0.96	0.97	0.97	1.0	1.2
n-Hexanol	0.44	0.48	0.51	0.53	0.52	1.5

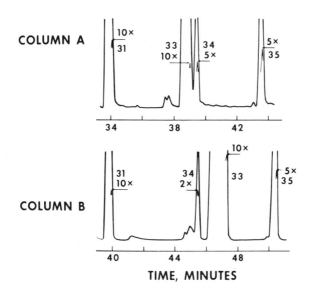

FIG. 3-6   Comparison of two "identical" columns, From
            Ref. (32).

Each column must be characterized very carefully for its specific
selectivities.

Not only does adsorption change selectivity, it also changes the
band broadening drastically. Figure 3-7 shows a separation of the
test compounds, chromatogram A, and a hydrocarbon mixture,
chromatogram B. If this column were judged only on separation of
hydrocarbons, it would be acceptable. However, if judged on its performance with n-hexanol (peak "a" in chromatogram A), then this column would be unacceptable. In gas chromatography-mass spectrometry (GC-MS) work used in aroma research (see Chapter 6), the tailing
of n-hexanol would cause its spectrum to persist as background for
many minutes. This high background would obscure the fragmentation
patterns of subsequently eluted compounds. Figure 3-8 shows how a
surface-active material, nonyl phenoxyethylene ethanol, Igepal
CO-880, lessens the amount of adsorption without perceptibly changing

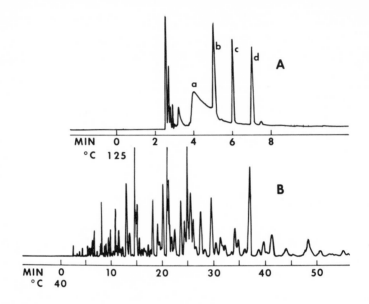

FIG. 3-7   Comparison of chromatograms of oxygenated compounds versus hydrocarbons. Column: 75-ft,
0.01-in.-i.d. open tubular column coated with
Apiezon C, no surface-active material added.
Chromatogram A: peaks a) n̲-hexanol, b) n̲-octanol,
c) n̲-decane, d) limonene. Chromatogram B:
hydrocarbon mixture. From Ref. (18).

RRT or selectivity of the liquid phase. Comparison of chromatogram
B with chromatogram A shows that increasing the amount of surface-
active material from 5 to 10% (w/w with respect to the stationary
liquid) reduces tailing of n̲-hexanol. Also, there is less band
spreading. More than 10% of surface-active material changes RRT
(18). James and Martin (12) used stearic acid with silicone oil for the
separation of volatile fatty acids. A subsequent paper (20) described
the use of sodium hydroxide to lessen adsorption of ammonia and
methyl amines. Averill (21) first used surface-active material to
decrease adsorption, and thus tailing, with open tubular columns. It
is readily seen in Figs. 3-7 and 3-8 that adsorption must be minimized
for separations in aroma research.

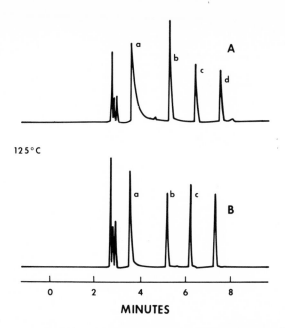

125°C

MINUTES

FIG. 3-8   Comparison of chromatograms of oxygenated compounds with addition of Igepal to Apiezon C. Column: 75-ft, 0.01-in.-i.d. Chromatogram A, 1% Igepal, 99% Apiezon C; Chromatogram B, 5% Igepal, 95% Apiezon C. Peaks labeled as in Fig. 3-7. From Ref. (18).

Tightness of bands, selectivity, and their contribution to resolution have been discussed. In aroma research, many investigators use the GC-MS combination. Therefore, separations must be accomplished with liquid phases which do not bleed much. Table 3-6 lists liquid phases which contribute little bleed into a mass spectrometer even at column temperatures up to 200° C. Of these, methyl silicone oils (for example, OV-101) are most used.

III.   COLUMNS

A.   Packed

Packed columns are the most commonly used because of their availability. They are inexpensive, rugged, and have good capacity. These columns should be used whenever possible.

TABLE 3-6

Relative Retention Times (Open Tubular)

	L (purified)	OV-101	OV-17	OV-25	OV-225	20M
n-Decane	1.0	1.0	1.0	1.0	1.0	1.0
Limonene	1.1	1.3	3.0	2.1	2.6	3.0
n-Octanal	0.58	0.96	2.0	2.2	4.9	4.2
n-Amyl acetate	0.38	0.46	1.1	1.2	2.2	2.2
n-Hexanol	0.22	0.37	0.9	1.1	3.3	4.5

Ottenstein (22) discussed various column support materials in detail. Porous polymer beads (23) are useful for separation of permanent gases and highly polar compounds but are not of general use in aroma research. Glass beads have not been known to yield high efficiency columns. However, textured glass beads are commercially available (24), and efficiencies up to 1000 TP/ft are claimed. These glass beads in glass or Teflon tubing could be useful for the separation of compounds which decompose on contact with metal.

B. Support-Coated Open Tubular

Support-coated open tubular (SCOT) columns were first described by Halasz and Horvath (25). A layer of very fine particles of a support material (diatomaceous earth) coated with a liquid phase is deposited on the inside wall of the column tubing. This addition of the fine particles permits a larger amount of liquid phase per given length of column than that on a wall-coated open tubular column originally proposed by Golay (26). SCOT columns can accept a relatively large sample charge. Although sample volumes of about 0.1-0.2 µl can be injected directly into these columns, the recommended sample volume for high efficiency is in the range of 0.01-0.05 μl. These sample volumes are not very much larger than those tolerated by 0.02-in.-i.d. wall-coated open tubular columns, and larger volumes can be applied to 0.03-in.-i.d. columns.

The SCOT columns give very good results with liquid phases which cover active sites, such as Carbowax, but there is a danger of adsorption with liquid phases like silicone and hydrocarbon oils. These oils do not seem to completely cover the larger area exposed by very fine particles of support material. These columns (27) provide the best efficiency of those commercially available, with the exception of standard wall-coated open tubular columns.

C. Wall-Coated Open Tubular

Wall-coated open tubular columns are not extensively used in aroma research, even though the first application of temperature programming of such columns was in this field (28). Some applications are listed by Harris and Habgood (8) and Ettre (7), but the full usefulness of this type of column has not been exploited. Therefore, their preparation and use will be discussed in some detail.

Schwartz and Brasseaux (29) point out that a special grade of tubing (30) is necessary for good columns. For most purposes, 0.0625-in.-o.d., with 0.01-, 0.02-, and 0.03-in.-i.d., tubing is used so that the same fittings can be used for all columns. For GC-MS work with no molecular separator interface, convenient lengths are 200 ft of 0.01-in.-i.d., 500 ft of 0.02-in.-i.d., and 1000 ft of 0.03-in.-i.d. These lengths of tubing give about 25 psi pressure drop across each column. Therefore, at 25 psi gauge pressure, the average linear velocity for a maximum number of TP is obtained with the detector end of the column at atmospheric pressure. When the column is connected to a mass spectrometer, a gauge pressure of 10 psi at the injector end yields an average linear velocity comparable to that obtained with the column outlet connected to a FID. Therefore, unless lengths are

otherwise specified, columns should be understood to be 200 ft X
0.01-in.-i.d., 500 ft X 0.02-in.-i.d., and 1000 ft X 0.03-in.-i.d.

For consistently good results, even new tubing should be thoroughly
washed (18). A reservoir (a stainless steel gas-sampling cylinder)
can be used for cleaning and coating columns. A 150-ml capacity is
convenient for cleaning; a 30-ml capacity, for coating. Cleaning
solvents and solutions are pushed through with nitrogen at 1000 psi or
more. All parts should be of stainless steel so that concentrated
nitric acid can be used. Four to five 150-ml portions of chloroform,
acetone, water, concentrated nitric acid, water, concentrated ammonia,
water, acetone, and chloroform are pushed through. If tubing has
been used for a prolonged time, it may be necessary to use more
nitric acid or ammonia to remove polymeric material which will not
come out with chloroform or acetone. A hot solution of 10% KOH in
methanol is sometimes useful to remove some materials which persist.
Unless the tubing is thoroughly cleaned of this semisolid material, a
good column cannot be prepared because of band spreading resulting
from resistance to mass transfer in the liquid phase. Deterioration
of the liquid phase due to depolymerization to smaller fragments is seen
as a rise in the base line with FID or as background in a mass
spectrometer. Some deterioration of the liquid phase results from
continued polymerization; this is especially evident with polyethylene
glycols such as Carbowax. Although the chloroform, acetone, and
water washings are clear, the nitric acid wash will carry out much
black material. The nitric acid wash must be continued until complete
removal. This procedure may seem tedious and time-consuming, but
it is a good investment toward making good columns. If the tubing is
cleaned thoroughly, it can be used over and over. In fact, older
tubings seem to yield better columns. The reason may be that nitric
acid treatment makes the inside surfaces smoother or simply that
coating techniques improve with experience. Certainly no deterioration in performance is observed unless the tubing has been abused
with extremely high temperatures or has been exposed to halogen

acids, which will attack stainless steel. All halogenated compounds should be avoided because they corrode stainless steel at elevated temperatures.

The cleaning and coating systems can be installed in a hood, so that the nitric acid, ammonia, and solvent fumes will not permeate the laboratory. The cleaning procedure need not take much experimental time because once the reservoir is filled and the nitrogen is turned on, the experimenter is free to continue other work.

The dynamic coating method was first described by Dijkstra and de Goey (31). The liquid film thickness is controlled by the concentration of the liquid phase in the solvent and by solvent evaporation rate. Evaporation rate is controlled by the pressure of the nitrogen. Columns with larger numbers of TP are obtained with solutions made with oxygenated solvents (32). Addition of a surface-active material not only decreases tailing of oxygenated compounds but also yields tighter bands even with hydrocarbons (18). Presumably, the surface-active material permits the liquid phase to wet the wall more evenly. For silicone oils, a 5% (w/w) solution is made with ether and acetone, 1:1, with 0.25% of Igepal added. For hydrocarbon liquid phases, such as apiezon L, a 5% solution is made with chloroform and ether, 1:1. For polyethylene glycols, such as Carbowax 20M, a 2% solution is made with chloroform, acetone, and ether, 1:1:1. Five or six portions of 10 ml for 0.03-in.-i.d., 5 ml for 0.02-in.-i.d., and 1 ml for 0.01-in.-i.d. tubing are pushed through at about 50 psi. After coating, the columns are left overnight at room temperature with the nitrogen at 50 psi. The columns are then conditioned for about a week at 175-200° C. The nitrogen flow is reversed during conditioning; i.e., the column is coated from the injector end but should be conditioned with the nitrogen introduced at the detector end. A purifier should be used to reduce the oxygen in the nitrogen to as low a concentration as possible. This is necessary to reduce the amount of oxidation of liquid phase during prolonged conditioning. If columns are conditioned for a week or more, little bleeding will be observed, even at 200° C. This conditioning is

imperative for GC-MS work. After conditioning, columns are placed
in a chromatograph and are characterized as to TP and selectivities.
Occasionally, columns should be reconditioned at least overnight at
200° C to remove any high molecular weight compounds that may have
been injected.

IV. SUMMARY

The successful culmination in GC is in achieving the desired separation, whether for detection or for isolation and identification. Table
3-7 summarizes sample load per peak possible with open tubular and
packed columns of various sizes. Of course, changing the temperature
changes the partitioning, and therefore the amounts that are tolerated
change. However, these values give a guideline to the amounts tolerated by different columns. A few overloads will ruin a 0.01-in.-i.d.
column, while 0.03-in.-i.d. columns survive overloading almost as
well as packed columns. If clean samples are injected, and if load
limits are observed, thousands of separations can be made with little
deterioration. Continuous overloading will strip any column. Large
overloads present a serious problem in GC-MS runs because the liquid phase is stripped into the mass spectrometer where it contributes
to background.

The 0.01-in. columns with an FID are useful for separation and detection of small quantities. They can be used to determine how many
components there are in a mixture or how pure a fraction may be.
Because the column capacity is so small, a stream-splitter injector
system is necessary. Thus, the 0.01-in.-i.d. column ofter requires
more sample than the 0.02- and 0.03-in.-i.d. columns, which can
utilize all of the sample. Even though the maximum sample load per
peak is about 5 μg for the 0.01-in.-i.d. columns, these columns
have been used effectively in the GC-MS combination because of the
sensitivity of the mass spectrometers. If the inlet pressure is adjusted to give the same average linear velocity to the mass spectrometer as to a FID, no loss of efficiency is observed (33). The low

TABLE 3-7

Sample Load Per Peak

	Column i.d., in.	Load, mg
Open tubular	0.01	0.005
	0.02	0.025
	0.03	0.25
SCOT	0.02	0.050
Packed	0.04	0.25
	0.1	0.5
	0.2	1.0
	0.5	50

volume flow rate of these 0.01-in.-i.d. columns permits them to
be connected directly to mass spectrometers.

The 0.02-in.-i.d. columns, in practice, have the highest resolution, with capacities of 25-50 μg/peak. These are very desirable for
use with GC-MS combinations. The 0.03-in.-i.d. columns have almost as much resolution as the 0.02-in.-i.d. columns and they have
capacities approaching those of 1/8-in.-o.d., 0.1-in.-i.d. packed
columns (about 0.25 mg/peak). The 0.02-in. columns are limited
to FID and MS work, but the 0.03-in. columns have enough capacity
to employ microthermistor detectors. With nondestructive detectors,
samples purified with 100,000-TP columns can be isolated for spectral or
sensory analyses. The high resolution and capacity of the 0.02- and
0.03-in. open tubular columns make them extremely valuable and
generally useful for various aspects of aroma research.

The capacity and resolution of SCOT columns place these between small bore open tubular and packed columns. These columns have good resolution for short columns, but they do not have the resolution or capacity of long 0.03-in.-i.d. open tubular columns.

The capacity of packed columns permits separation of milligram quantities of fractions which contain just a few components. It is interesting to note that 0.04-in.-i.d. packed columns have no more capacity than 0.03-in.-i.d. open tubular columns. The pressure drop across a packed column prohibits construction of columns with a large number of TP. A 0.1-in.-i.d., high efficiency packed column does not have much more capacity than the 0.03-in. open tubular column. However, it must be admitted that the packed column can tolerate a bigger overload. Therefore, the 1/4-in.-o.d., 0.2-in.-i.d. packed column is the next practical size. A 100-ft column may have 10,000-50,000 TP and a capacity of 1 mg/peak (32). Most packed columns have only several thousand TP. Preparative columns, 0.5-in. i.d., can be made with capacities of 50 mg/peak. Good separations can be achieved with columns having only a few thousand TP if enough material is available for numerous repeated separations utilizing different selectivities.

Figure 3-9 illustrates the type of separation necessary to study aroma problems in detail. Buttery et al. (34,35) have shown that different varieties of hops can be identified by their chromatograms. These chromatograms were obtained with a 150-ft, 0.01-in.-i.d., open tubular column coated with methyl silicone oil SF 96(50) liquid phase, with Igepal CO-880 added as tail reducer. Even though the column was temperature programmed and each separation took over 4 hr, the retention times were reproducible enough that changes in amounts of constituents in the different varieties could be followed.

Figure 3-10 shows a separation of coffee aroma complex (36). This mixture is the most complex known in aroma research, with over 400 constituents shown. Separation was achieved with a glass open

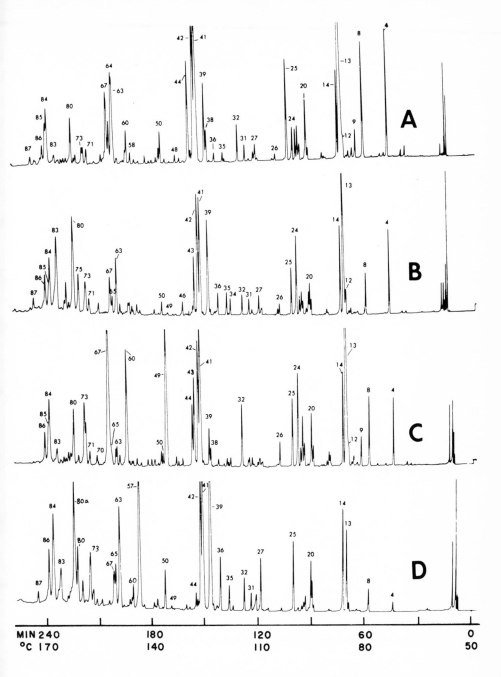

FIG. 3-9 Programmed temperature capillary column, 0.01-in.
i.d. separations of oxygenated compounds found in
four different varieties of hops. From Ref. (34).

101

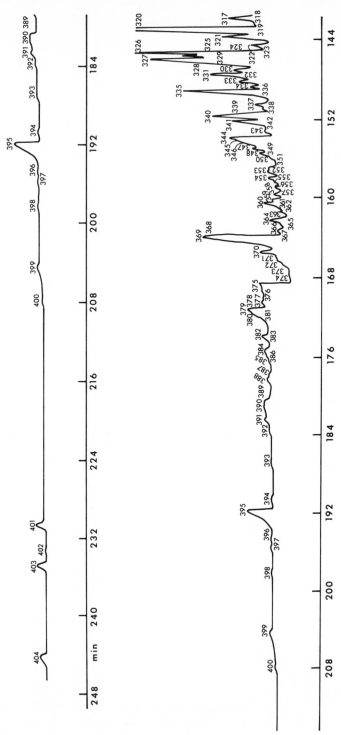

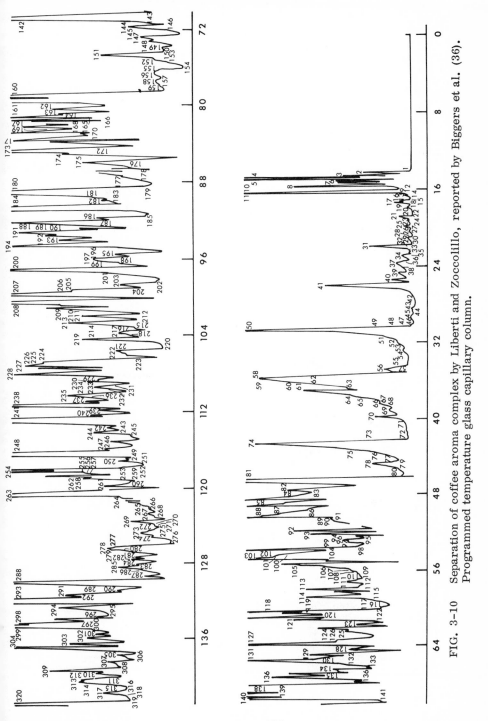

FIG. 3-10   Separation of coffee aroma complex by Liberti and Zoccolillo, reported by Biggers et al. (36). Programmed temperature glass capillary column.

103

tubular column. Some flavor compounds are very labile and decompose on contact with hot metal. If compounds are not destroyed, adsorption problems remain. Therefore, all-glass systems should be more widely used in flavor research.

In aroma research, sensorially important compounds must be separated from incredibly complex mixtures isolable only in very small amounts, without damage to the sensory properties. Resolution has been briefly discussed with respect to tightness of bands and selectivities as well as some of the parameters important in each. Open tubular, support-coated tubular, and packed columns have been compared. Although more progress must be made in various areas of GC, many problems in aroma research can be solved with available techniques and instruments.

REFERENCES

1. D. G. Guadagni, in Correlation of Subjective-Objective Methods in the Study of Odors and Taste, American Society for Testing and Materials, Philadelphia 1968, p. 36.

2. A. J. Haagen-Smit, in Flavor Research and Food Acceptance, Reinhold, New York, 1958, p. 369.

3. J. W. Ralls, paper presented at the Western Experiment Station Collaborators Conference, WURDD, USDA, Albany, Calif., 1962.

4. S. Dal Nogare and R. S. Juvet, Jr., Gas-Liquid Chromatography, Wiley (Interscience), New York, 1962.

5. A. B. Littlewood, Gas Chromatography, Academic, New York, 1962.

6. H. Purnell, Gas Chromatography, Wiley, New York, 1962.

7. L. S. Ettre, Open Tubular Columns in Gas Chromatography, Plenum, New York, 1965.

8. W. E. Harris and H. W. Habgood, Programmed Temperature Gas Chromatography, Wiley, New York, 1966.

9. The Practice of Gas Chromatography (L. S. Ettre and A. Zlatkis, eds.), Wiley (Interscience), New York, 1967.

10. O. E. Schupp, III, Gas Chromatography, Technique of Organic Chemistry (E. S. Perry and A. Weissberger, eds.), Vol. XIII, Wiley (Interscience), New York 1968.

11. H. G. Cassidy, Fundamentals of Chromatography, Wiley (Interscience), New York, 1957.

12. A. T. James and A. J. P. Martin, Biochem. J., 50, 679 (1952).

13. E. Gil-Av, B. Feibush, and R. Charles-Sigler, Tetrahedron Letters, No. 10, 1009 (1966).

14. S. Dal Nogare, Anal. Chem., 37, 1450 (1965).

15. R. G. Buttery, R. E. Lundin, and L. Ling, J. Agr. Food Chem., 15, 58 (1967).

16. L. Rohrschneider, Z. Anal. Chem., 170, 256 (1959).

17. T. R. Mon, R. R. Forrey, and R. Teranishi, J. Gas Chromatog., 4, 176 (1966).

18. T. R. Mon, R. R. Forrey, and R. Teranishi, J. Gas Chromatog., 5, 497 (1967).

19. R. A. Flath, D. R. Black, D. G. Guadagni, W. H. McFadden, and T. H. Schultz, J. Agr. Food Chem., 15, 29 (1967).

20. A. T. James, A. J. P. Martin, and G. H. Smith, Biochem. J., 52, 238 (1952).

21. W. Averill, in Gas Chromatography (N. Brenner, J. E. Callen, and M. D. Weiss, eds.), Academic, New York, 1962, p. 1.

22. D. M. Ottenstein, J. Gas Chromatog., 1(4), 11 (1963).

23. O. L. Hollis, Anal. Chem., 38, 309 (1966).

24. Corning Glass Works, Corning, N. Y.

25. I. Halasz and C. Horvath, Anal. Chem., 35, 499 (1963).

26. M. J. E. Golay, in Gas Chromatography 1960 (R. P. W. Scott, ed.), Butterworths, Washington, D. C., 1960, p. 139.

27.   Perkin-Elmer Corp. , Norwalk, Conn.

28.   R. Teranishi, C. C. Nimmo, and J. Corse, Anal. Chem. , 32, 1384 (1960).

29.   R. D. Schwartz and D. J. Brasseaux, Anal. Chem., 35, 1374 (1963).

30.   Handy and Harmon Tube Co. , Norristown, Pa.

31.   G. Dijkstra and J. de Goey, in Gas Chromatography 1958 (D. H. Desty, ed.), Butterworths, London, 1958, p. 56.

32.   T. R. Mon, USDA, Albany, Calif. , private communication, 1969.

33.   R. Teranishi, R. G. Buttery, W. H. McFadden, T. R. Mon, and J. Wasserman, Anal. Chem., 36, 1509 (1964).

34.   R. G. Buttery, D. R. Black, D. G. Guadagni, and M. P. Kealy, Am. Soc. Brewing Chemists' Proc. , 1965, p. 103.

35.   R. G. Buttery and L. A. Ling, J. Agr. Food Chem. , 15, 531 (1967).

36.   R. E. Biggers, J. J. Hilton, and M. A. Gianturco, in Advances in Chromatography (A. Zlatkis, ed.), Preston Technical Abstracts, Evanston, Ill. , 1969, p. 238.

Chapter 4

SPECIAL APPLICATION OF GAS CHROMATOGRAPHY

I. VAPOR ANALYSIS

Evaluation of quality is one of the greatest difficulties in aroma research. The ideal solution would be a rapid, easy instrumental analysis giving a detailed, quantitative picture of the components present, to replace the time-consuming, expensive taste-panel tests. In an attempt to achieve such analyses, GC instruments with a low limit of detection have been built for head space, or direct vapor, analyses. However, we must understand not only what we can do with these instruments but also what their limitations are. Table 3-1 in

Chapter 3 shows some odor threshold values. The best instruments with FID can produce chromatograms of samples containing parts per billion of organic material; therefore, our instruments can barely detect n̲-amyl acetate or dimethyl sulfide at threshold concentrations. From this table of threshold values, we see that even the theoretical limit of an FID is far from being sensitive enough to detect such compounds as 2-methoxy-3-isobutylpyrazine or β-ionone at threshold concentrations. Thus, it is imperative that we operate an instrument as close as possible to its theoretical limit. To do so, we must consider the experimental parameters which are important.

A.   Instrumentation

A hypothetical GC peak is shown in Fig. 4-1 which illustrates signal, noise, and drift. Superimposed on this peak is a second peak which is drawn without noise or drift and represents a tighter, more concentrated band. The areas are about equal, but it is obvious that the signal-to-noise ratio is much more favorable in the case in which the tighter band is delivered to the detector. This example shows that the signal must be maximized by detector sensitivity parameters.

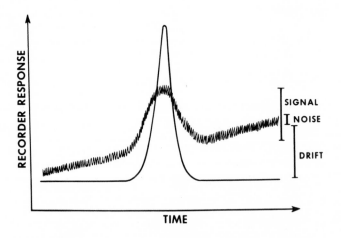

FIG. 4-1   Hypothetical GC peaks.

The signal-to-noise ratio must also be improved by utilizing the highest efficiency columns available.

1. Sensitivity

First, for maximum sensitivity with an FID, maximum signal must be obtained for a given amount of organic material going through the detector. Factors influencing the performance of an FID have been summarized by Sternberg et al. (1). For optimum signal, a ratio of nitrogen to hydrogen of 1:1 must be maintained. For a given burner tip orifice size, there is an optimum flow. This flow rate should be maintained whether a small diameter open tubular column or a packed column is used. If the flame burns too low because of small flow rate, the tip of the burner assembly will overheat, and thermionic emission will cause a very high background noise level. Figure 4-2 shows side and top views of a hydrogen burner assembly. Effluents from the column, E, are mixed with hydrogen and nitrogen (introduced through a "T" at D), introduced to the burner together at A, and burned at the

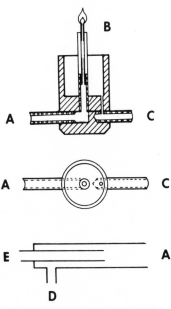

FIG. 4-2   Burner assembly.

tip, B. Air is introduced to ensure complete combustion. Hydrogen
and nitrogen can be premixed before introduction at D to maintain the
1:1 ratio. This ratio can be maintained by adjusting the pressure of
each to a restrictor inserted between the pressure regulator and the
point of introduction, D. With this assembly, open tubular and packed
columns can both be accommodated without any change in plumbing to
the burner assembly. Moreover, since the end of the column, E, is
in the stream of nitrogen and hydrogen, the effluent is swept to the
burner tip in a greater flow than that with just carrier gas. Severe
band spreading can be caused by small pockets in which there can be
eddying with low flow rates. Such eddying effects are minimized with
addition of the gases at D. Moreover, with the organic material diluted
by hydrogen and nitrogen, the dew point is lowered and the operating
temperature can be reduced. Signal from an FID depends on mass flow
rate of oxidizable, or ionizable, material going through the flame, not
on its concentration. The nitrogen increases the signal-to-noise ratio
with the FID. Dilution of material going through a thermal conductivity
detector (TCD) would obviously decrease the signal.

Figure 4-3 shows a GC flow diagram (2). The following factors can
contribute to drift: bleed from the liquid phase, high molecular weight
compounds eluting days after injection, impurities and flow rate
changes in the gases used, back-diffusion of organic vapors from the
room atmosphere into the detectors, changes in the voltage, resistors

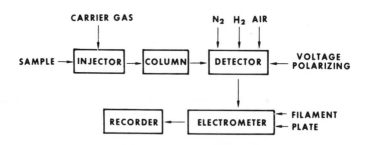

FIG. 4-3   GC flow diagram (2).

changing in their values due to temperature changes in the detector
and in the electrometer, etc. Of these, the most difficult to minimize
is the bleeding from the liquid phase. A compromise must be made in
the choice of phase. Low molecular weight material of low viscosity
is required for high efficiency columns (see Chapter 3). High molecular weight provides low vapor pressure and less column bleed.
Bleed from a liquid phase can be reduced by conditioning at higher
temperatures than those used in normal operation. If care is taken to
reduce trace amounts of oxygen in the carrier gas used during conditioning, the columns can be conditioned for a week or so without
oxidative damage to the liquid phase. Injection of any acidic material
may start depolymerization of the liquid. Once this reaction starts,
the column must be replaced. For minimum drift, columns should be
cleared of slowly eluting materials by keeping them at elevated
temperatures at least overnight. Impurities in the carrier gas, hydrogen, nitrogen, and air can be removed by molecular sieves, alumina,
silica, or activated charcoal. Prescott and Wise (3) have shown that
clean gases help to achieve more accurate results and better detector
sensitivity. Drift caused by fluctuations in the gas flow rates can be
lessened by using restrictors before and after surge tanks. Restrictors to surge tanks should have greater flow rates than those after the
tanks. Capacity of the tanks should be large enough so that the time required to fill the tank is greater than the cycle period of the pressure
regulator. Line voltage surges can be avoided by using batteries for
polarizing, filament, and plate voltages. However, solid state power
supplies which are far superior to the vacuum tube type are now commercially available. Resistance changes can be minimized by the use
of the McWilliam and Dewar dual flame and differential electrometer
system (4).

Vibrations cause noise in vacuum tube electrometers. Current is
generated as the filaments in the tube move in the magnetic field.
Such noise can be greatly reduced by shock mounting the tube and the
entire chassis. If the high frequency vibrations are reduced, the

noise level is reduced considerably. The motor used to circulate the
air in the oven is the primary source of vibration in a GC. Such noise
difficulties can be virtually eliminated by using all-solid-state electrometers. Figure 4-4 shows such a system (5). Components can be
purchased and easily assembled. The electrometers shown in Fig.
4-4 can be used independently or in a differential mode. Some commercial instruments now have solid state electrometers.

The distance between the burner and the electrometers should be
as small as possible to reduce pickup of extraneous noise. The best
input resistance selector switch should be used. All line voltage wires
and power supplies should be kept as far away as possible from the
electrometer area. Most electrometers are more than adequate for
FID systems, but often the performance of the electrometer, and
therefore of the entire GC unit, is jeopardized because of the position
of the components with respect to one another.

Generally, the FID is the most sensitive detector, and it is the
most widely used. However, other types of detectors which are more
often used in pesticide and herbicide analyses should not be ignored.
Perhaps specific detectors for halogens or phosphorus may not be of
particular value in flavor research, but certainly detectors specific
for nitrogen and sulfur would be of interest (see Chapter 8 for nitrogen
and sulfur compounds of flavor importance). Although signals from a
GC detector cannot be taken as proof positive that nitrogen or sulfur
compounds are present in a mixture, certainly such indications can be
of great value. Even if a mass spectrometer is accessible, there are
occasions in which it may be desirable to quickly scan mixtures for
indications of nitrogen or sulfur compounds.

2. Limit of Detection

Sensitivity concerns only the detector. Limit of detection concerns
the entire analytical system. Once detector parameters are optimized,
then attention must be focused on delivering each separated organic
material to the detector in as concentrated a band as possible. To

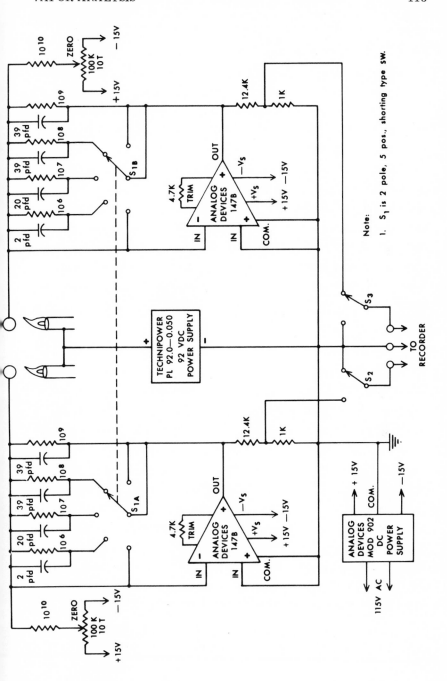

FIG. 4-4   Solid state electrometers.

begin with, the sample must be introduced in as tight a band as possible. There must be no large volumes in which any eddying and band spreading can occur in the system from the injector to the detector. The greatest improvement in the limit of detection can be achieved with use of very high efficiency columns, i.e., columns which deliver separated material in the tightest bands. Use of open tubular columns not only gives greater resolution of separation but also lowers the limit of detection. Nawar and Fagerson (6) obtained vapor chromatograms with 0.01-in.-i.d. open tubular columns with sample enrichment techniques. Flath et al. (7) used precolumn trapping and a valving system for preliminary concentration of condensables which were separated with a 0.03-in.-i.d. open tubular column. The ruggedness and sample size tolerance of the 0.03-in.-i.d. columns make them more practical and useful than the 0.01-in.-i.d. columns.

B. Sample Preparation

Although direct vapor sampling can give some interesting results, often not enough material is available in a 10- or 20-cc sample, and enrichment can provide much better head space analyses. MacKay (8) showed that trace amounts can be trapped in concentration tubes filled with column packing. Hornstein and Crowe (9) showed that a trapping coil cooled in liquid nitrogen would collect volatiles swept from the food product with nitrogen. Morgan and Day (10) and Arnold and Lindsay (11) have used this method to collect volatiles from dairy products. Nawar and Fagerson (12) showed that recycling vapors through traps enriched the sample considerably. Dravnieks and Krotoszynski (13) showed that traps at room temperature filled with solid support coated with liquid phase will collect organic vapors while letting water vapor through. Also, Wientjes (14) has shown that the organic vapor concentration can be increased by increasing the sugar concentration in aqueous solutions.

C. Examples

Although the limit of detection of GC instruments is still orders of magnitude higher than that of the nose, some practical applications can be made in surveying products to give indications of differences in quality. Buttery and Teranishi (15) have shown that direct injection of food vapors can be used to follow the development of oxidative rancidity and browning. Figure 4-5 shows chromatograms obtained by direct injection of vapor above reconstituted dehydrated potato granules stored for different periods in oxygen. Although hexanal is not mainly responsible for the characteristic aroma of rancidity, the increase in the concentration of hexanal, the biggest peak, can be related to the development of rancidity (15).

Figure 4-6 (16) shows chromatograms which illustrate the difference between two commercial apple juices. Chromatogram B shows that

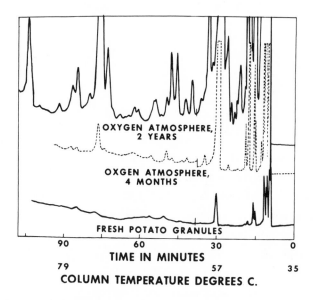

FIG. 4-5   Direct vapor analyses of reconstituted potato granules
            stored under oxygen atmosphere (15).

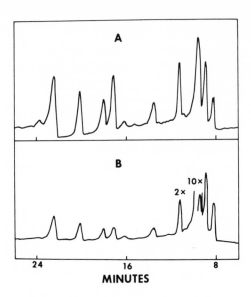

FIG. 4-6   Direct vapor analyses of apple juices, Brand A and
            Brand B.

this product has a much higher concentration of ethanol (10 x peak)
than product A. Although there was no great difference in quality
between these two products, these chromatograms show that such head
space analyses can pick up small differences in the amount of ferment-
ation that has occurred.

McCarthy et al. (17) have shown qualitative and quantitative correlations between the development of flavor notes and chemical
analyses for volatiles by GC with data obtained on two banana varieties.
Figure 4-7 shows differences in concentration of constituents in the
head space as the fruit ripens. Correlations found by these workers
substantiate the potential use of GC in routine evaluation of fruit
quality. As more compositional results are correlated with sensory
qualities, and as limits of detection and resolution of separations are
improved, such data will become even more valuable.

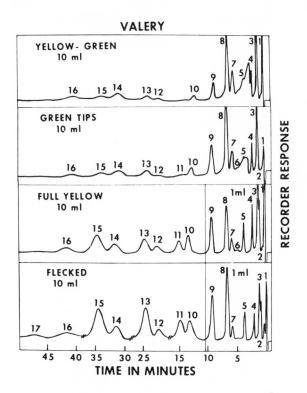

FIG. 4-7   Head space analyses of ripening banana (17).

Nawar and Fagerson (6) and Heins et al. (18) have used 0.01-in.-
i.d. open tubular columns for direct vapor analyses. Flath et al. (7)
used 0.03-in.-i.d. open tubular columns. Flow rates and sample
capacity of the larger bore columns make them much more practical
for direct vapor analyses. Figure 4-8 shows a valving system which
is used to condense the volatiles introduced via the injector. This system has the advantage that there is very little back pressure since the
noncondensables are vented. On turning the valve, the trap is brought
into the column system, and the volatiles can be introduced onto the
column as a very sharp band (7).

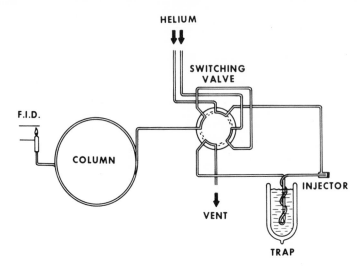

FIG. 4-8   Valve system for trapping samples for open tubular
            column vapor analyses (7).

Figure 4-9 shows chromatograms of four brands of apple juice, obtained with a 0.03-in.-i.d. open tubular column (19). When the
chromatograms in Fig. 4-9 are compared with those of Fig. 4-6, it is
obvious that resolution of separation is improved. The limit of detection, because of the much sharper peaks, has been lowered markedly.
Brand A has the least amount of ethanol (peak at 19 min). Brands B,
C, and D have so much ethanol that tailing is observed even though the
carrier gas is partially saturated with water. That Brand A was
superior was obvious merely by sniffing. No statistical analysis of
taste panel results was necessary. The chromatograms show that
Brand A has over four times more butyl acetate (at 55 min) and over
10 times more 2-methylbutyl acetate (70 min) than the other brands.
The two compounds important in Delicious apple essence (20), trans-
2-hexenal and ethyl 2-methylbutyrate, elute at about 64 min. After a
glance at chromatogram D, it comes as no surprise that Brand D has
little aroma.

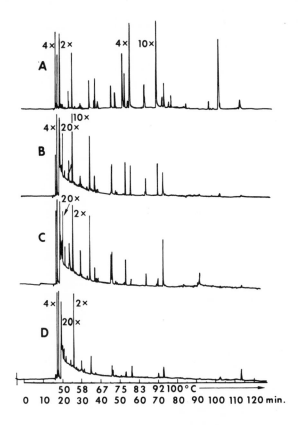

FIG. 4-9   Direct vapor analyses of four apple juice brands.
            Column: 1000-ft, 0.03-in.-i.d. open tubular
            coated with SF 96 (50) methyl silicone oil (19).

These examples illustrate how meaningful vapor analyses can be
after chemical structure identification and sensory evaluations have
been accomplished.

II. PROGRAMMING

A. Temperature

Harris and Habgood (21) have discussed the advantages of programming column temperatures. Almost all commercial instruments have

this capability. With this technique, fractions eluting hours after injection can be brought to the detector in sharp bands, especially with long open tubular columns; therefore, the limit of detection can be kept relatively constant throughout the range of compounds being analyzed. Delivering material in high concentration in the carrier gas is important in GC-MS analyses. A much better signal-to-noise mass spectrum is obtained if the material is delivered in several seconds to a fast scan mass spectrometer than if the same amount is delivered in several minutes.

It is generally accepted that the lower the temperature, the better the separation. Slow programming of column temperature permits separation at the lowest temperature possible. Various program rates must be tried to obtain the best separation. For example, there may be a reversal of retention with change of temperature (22) (see Fig. 4-10). Trans-2-hexenal (peak 1) elutes before ethyl 2-methylbutyrate (peak 2) from an open tubular column coated with methyl silicone

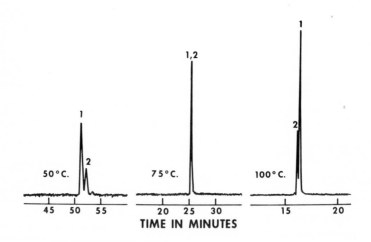

FIG. 4-10   Change in relative retention times with temperature.
Peak 1, trans-2-hexenal; peak 2, ethyl
2-methylbutyrate. Column: 650-ft, 0.03-in.-i.d.
open tubular coated with SF 96(50) methyl silicone
oil (22).

SF 96(50) at 50°C. At 75°C, these compounds elute together. At
100°C, ethyl 2-methylbutyrate elutes before trans-2-hexenal. For
these two compounds, lowering the column temperature from 100 to
75°C does not result in better separation, whereas the decrease from
75 to 50°C does. This example, with sensorially important compounds
(20), serves as a warning that various temperature program rates
must be tried to obtain the resolution desired.

B. Pressure

As the temperature of the column is raised, the pressure drop
across the column increases. If a constant linear velocity of carrier
gas is to be maintained, inlet pressure should be continuously
increased. Very few operators bother to maintain constant linear
velocity because slight deviations from optimum conditions do not result in serious band spreading. In fact, deliberate deviation from
optimum conditions, i.e., pressure programming far beyond the usual
linear velocities, has resulted in better separations in some cases
(23). Figure 4-11 shows chromatograms of orange oil fractions (24).
Chromatogram A shows a separation in which temperature is programmed at constant pressure. The program rate should have been
lower since the earlier peaks are crowded. Chromatogram B shows
an isothermal separation with the pressure programmed. The earlier
peaks are well separated, but the skewed peaks at the end show that
the temperature is too low for the higher molecular weight compounds.
Chromatogram C shows the best separation obtained. By pressure
programming first, then temperature programming, all the peaks,
early and late, are well resolved. Moreover, peaks not seen by either
type of programming alone are now detectable. Thus, the range that
can be surveyed with a given column has been extended.

One of the main reasons why pressure programming is not
extensively used is that the pressure will drift with the usual
pressure-programming devices, and elution times are not

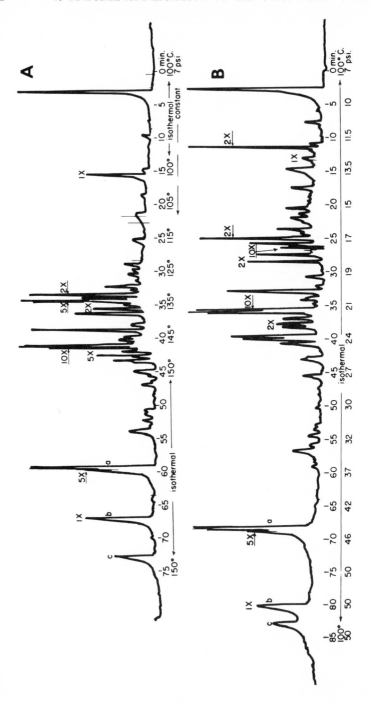

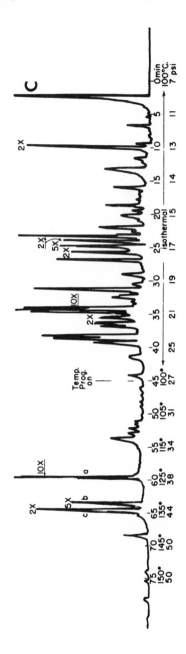

FIG. 4-11 Temperature and pressure programming. Chromatogram A, temperature programmed. Chromatogram B, pressure programmed. Chromatogram C, pressure and temperature programmed. Column: 75-ft, 0.01-in.-i.d., coated with SF 96(50) methyl silicone oil.

reproducible. It is hoped that this experimental difficulty will be eliminated.

Response from an FID does not vary much as the flow changes with pressure programming, especially if relatively large flows are maintained at the burner with hydrogen and nitrogen makeup gases and with a differential signal system with dual detectors. With thermal conductivity detectors, which are concentration dependent, response falls considerably as the flow is increased and solutes are diluted with a greater volume of carrier gas. Moreover, thermistor detectors are very flow sensitive, and the base line drifts considerably if a single column is used. With a dual-column system, little base line drift is observed if the two columns are closely matched. The second auxiliary column should maintain the same flow rate through the reference side as that through the detector side (24).

III. PREPARATIVE CHROMATOGRAPHY

Even 0.01-in.-i.d. open tubular columns may be considered preparative when used with MS. The 0.03-in. columns can be used with thermistor detectors for separation and isolation of milligram quantities for spectrometric analyses. Generally, however, preparative GC implies use of large diameter packed columns.

Usually, large diameter coiled packed columns provide low efficiency, i.e., only about 1000 theoretical plates (TP). One of the major contributions to band spreading in such columns results from different linear velocities in different areas in the column (25), caused by particles of varied density and size settling preferentially. For high efficiency packed columns, support particles should be carefully selected for small range of size and density (26). In spite of low TP values, these columns are very useful in preparing fractions with relatively few components. Moreover, if sufficient starting material

is available, multiple separations with liquid phases of different selectivities will often yield fractions of high purity.

Carle and Johns (27) have shown that beyond 1.5-cm i.d. efficiency
decreases rapidly. This decrease was verified by packing 1.25- and
2.0-cm tubing with portions from the same batch of carefully selected
and impregnated support material. The 1.25-cm column yielded about
three times as many TP per unit length as the 2.0-cm column. When
the packing was removed from the 2.0-cm column and was repacked
into a 1.25-cm tube, efficiency was identical to that of columns made
by packing 1.25-cm columns with freshly selected and impregnated
support material (28).

Columns (1.25-cm i.d., 10 m, with 1-m vertical U-tube configuration) packed with carefully selected and impregnated support material
(4% liquid phase on 80/100 mesh Chromosorb G) can have more than
5000 TP and a capacity of about 50 mg/peak (28). Figure 4-12 (29)
compares a preparative separation with one from a 0.01-in.-i.d. open
tubular column. Of course, the resolution with the 1.25-cm packed
column is not as good as that with the 0.01-in. open tubular column.
However, tens of milligrams are separated with the former, in con-
trast to the micrograms separated with the open tubular column.

Most commercially available preparative columns do not provide
more than 1000 TP. Some handle grams or tens of grams per fraction. Resolution in distillation is directly proportional to the number
of TP. Because of the different definition, resolution in GC is proportional to the square root of TP. Therefore, more effective
separations are achieved with a good distillation column (see Chapter
2) than with a low resolution preparative GC, if enough starting
material is available and if it is stable to prolonged exposure to
elevated temperatures. Distillation cannot be used in cases in which
the alcohol moieties of acetals, ketals, or esters exchange or when
some very heat-labile compounds such as allylic tertiary alcohols or

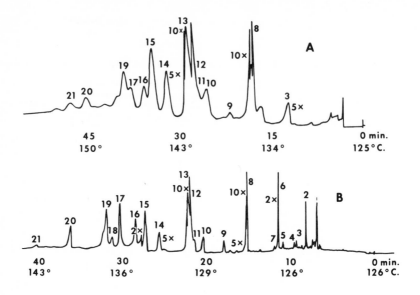

FIG. 4-12   Preparative column versus capillary column separations (29).   Chromatogram A column: 30-ft, 0.5-in.-
i.d., packed with 60/70 mesh Chromosorb G, impregnated with 5% SF 96(50) methyl silicone oil.
Chromatogram B column: 200-ft, 0.01-in.-i.d.,
coated with SF 96(50) methyl silicone oil.

their acetates are present. When properly used, low resolution preparative GC or distillation can greatly assist isolation of relatively
simple fractions for subsequent study.

   Fractions can be separated according to functional groups with
liquid-solid adsorption chromatography (see Chapter 2). Combination
of separation techniques will ultimately yield a pure fraction. Unfortunately, the aroma chemist seldom has sufficient material to
utilize all the powerful methods available.

   Great care must be exercised in the application of any separation
technique. At each step, sensory properties should be assessed.
Proportional amounts from all distillation fractions should be taken in

order to reconstitute the original mixture. The reconstituted material
should have the same sensory properties as the original. Similar
techniques should be applied to fractions from chromatographic operations. Often, important compounds are decomposed or irreversibly
adsorbed on the column, causing gross changes in sensory properties.
All fractions may be collected in a single vessel, and this material
should be compared to the original sample to check if sensory
properties are altered by the procedure.

IV. GAS CHROMATOGRAPHY-SENSORY EVALUATION

Each time a separation with a nondestructive detector is accomplished, the investigator may make a sensory evaluation of each fraction. Panel evaluation of GC effluents is discussed in Chapter 8.
Ryder (30) showed that many different odors can be detected from any
single injection of an isolate from a food product. Expert perfumers
(31) also use this method.

There are some obvious difficulties with sniffing the effluents from
a GC. Concentrations change rapidly, and it is well known that sensations vary with concentrations. Even with the highest resolution columns, there is no guarantee that each peak represents only one compound. Decomposition, not detected by a spectrometric method, may
change or mask the sensory character of a fraction. Desirable aromas
may be exhibited by mixtures. When their components are separated,
only an experienced flavorist may be able to relate their sensory
properties to the original aroma. Nevertheless, sniffing the GC
effluents can give some very interesting indications as to which fractions may be of sensory importance (see Chapter 8).

Various arrangements of GC equipment can permit "nasal
appraisals." One is to pass the GC effluent through a hot wire thermal
conductivity detector (TCD) or a thermistor TCD. The thermistor is
more sensitive than the hot wire TCD. Moreover, the thermistor is

glass-coated, dissipates about 1/10 the energy of a hot wire TCD, and operates at lower temperature. Thus, there is less chance for decomposition with a thermistor than with a hot wire TCD.

Another arrangement splits part of the column effluent to an FID. Because of the FID's greater sensitivity as compared to the TCD, only a tenth or a hundredth of the column effluent need reach the FID to give a lower limit of detection than that provided by a TCD. The advantage of this technique is that the effluent is not exposed to temperatures higher than that of the column. Often, detectors are kept 25-50° C higher than the column temperature to prevent any condensation. These high temperatures may cause decomposition of labile compounds. In view of temperature considerations, open tubular columns are more desirable than packed columns. For a given compound, an open tubular column can be operated 50° C cooler than a packed column with the same liquid phase. Thus, for the sniff test, the most desirable system is one which incorporates a 0.03-in. open tubular column with a split to an FID. The hot wire TCD should not be used because of high filament temperature, low sensitivity, and dead volume. A thermistor TCD should be kept as cool as possible without condensation. A particularly desirable feature in using a split to a FID is that the carrier gas can be partially saturated with water vapor (32). Water vapor reduces adsorption on the column and lessens irritation of the sniffer's nasal membranes caused by inhaling dry carrier gas.

V.   MISCELLANEOUS PROBLEMS

A.   Band Spreading

Band spreading can occur in large volumes in the injector or in the connection to the detector. A cold spot in the system can also cause serious band spreading. All connections from the injector to the column and from the column to the detector should be kept as short as possible to minimize dead volumes.

An injector with a replaceable insert is desirable. Small particles of silicone rubber septum break off each time a syringe needle is inserted. These particles may accumulate in the injector so that an appreciable amount of sample can be absorbed by the silicone rubber. Because of the time required for desorption from these particles, tailing is observed even with hydrocarbons. The number of TP may drop from over 100,000 to less than 10,000 simply because of silicone rubber particles in the injector (28). With systems having only a few thousand TP, such an effect is not noticeable. For systems having over 100,000 TP, it is imperative that the injector be kept scrupulously clean.

Figure 4-13 shows a commercial injector (33) with a low dead volume glass insert, D, which can be replaced easily and routinely after a few injections to ensure a clean injector. This system can also be used as a precolumn by using an insert, D', packed with some impregnated solid support material. Extra packed inserts can be prepared so that they can be changed after each injection. This technique not only ensures a clean injection system but also protects the column from contamination.

B. Decomposition

A dirty injector can cause chemical rearrangements. If acidic material is present, sabinene can be quantitatively rearranged to α- and γ-terpinene (28). Also, α-thujene will rearrange to the terpinenes under somewhat more drastic conditions. Conversion of cis- and trans-isomers can occur easily. Pure neral or geranial is converted to a 1:1 mixture of both compounds by a dirty injector or dirty stream splitter (28). If basic material is present, epimerization can occur.

Thermal decomposition can be catalyzed by acid or base, as in the case of acid dehydration of tertiary, allylic alcohols. Pyrolysis of some tertiary acetates occurs quantitatively at 175° C. Tertiary amyl acetate will survive GC purification if injected on a cool column which

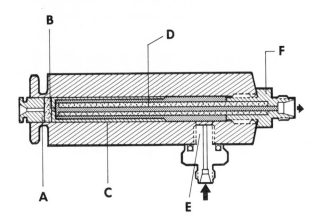

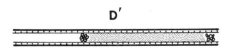

FIG. 4-13   Injector System with Replaceable Inserts. A,
Spring-loaded septum retainer and needle guide.
B, Septum. C, Glass liner. D, Glass vaporizer
tube. D', Glass vaporizer tube partially filled
with impregnated support material. E, Input
fitting. F, Column fitting.

is temperature programmed slowly. Some compounds, like methional,
react with metals, especially copper. If any trace of such metals
remains, reactive compounds will not survive GC separations. With
very labile compounds, all-glass or Teflon systems are suggested.
Glass beads seem to be the most inert support material.

GC is indispensable in flavor research. It must be used with extreme
care to achieve separation of labile compounds from complex mixtures.
The flavor chemist must assure retention of sensory properties at
every step.

REFERENCES

1. J. C. Sternberg, W. S. Gallaway, and D. T. L. Jones, in Gas Chromatography (N. Brenner, J. E. Callen, and M. D. Weiss, eds.), Academic, New York, 1962, p. 231.

2. R. Teranishi, R. G. Buttery, R. E. Lundin, W. H. McFadden, and T. R. Mon, Am. Soc. Brewing Chemists Proc., 1963, p. 52.

3. B. O. Prescott and H. L. Wise, J. Gas Chromatog., $\underline{4}$(2), 80 (1966).

4. R. Teranishi, R. G. Buttery, and R. E. Lundin, Anal. Chem., $\underline{34}$, 1033 (1962).

5. H. Gill, Analog Dialogue, $\underline{1}$(2), 1 (1967).

6. W. W. Nawar and I. S. Fagerson, Food Technol., $\underline{16}$(11), 107 (1962).

7. R. A. Flath, R. R. Forrey, and R. Teranishi, J. Food Sci., $\underline{34}$, 382 (1969).

8. D. A. M. MacKay, in Gas Chromatography-Edinburgh 1960 (R. P. W. Scott, ed.), Butterworths, London, 1960, p. 357.

9. I. Hornstein and P. F. Crowe, Anal. Chem., $\underline{34}$, 1354 (1962).

10. M. E. Morgan and E. A. Day, J. Dairy Sci., $\underline{48}$, 1382 (1965).

11. R. G. Arnold and R. C. Lindsay, J. Dairy Sci., $\underline{51}$, 224 (1968).

12. W. W. Nawar and I. S. Fagerson, Anal. Chem., $\underline{32}$, 1534 (1960).

13. A. Dravnieks and B. K. Krotoszynski, J. Gas Chromatog., $\underline{6}$, 144 (1968).

14. A. G. Wientjes, J. Food Sci., $\underline{33}$, 1 (1968).

15. R. G. Buttery and R. Teranishi, J. Agr. Food Chem., $\underline{11}$, 504 (1963).

16. R. Teranishi, USDA, Albany, Calif., unpublished results.

17. A. I. McCarthy, J. K. Palmer, C. P. Shaw, and E. E. Anderson, J. Food Sci., $\underline{28}$, 379 (1963).

18. J. T. Heins, H. Maarse, M. C. ten Noever de Brauw, and C. Weurman, J. Gas Chromatog., $\underline{4}$, 395 (1966).

19.   R. R. Forrey, USDA, Albany, Calif., unpublished results.

20.   R. A. Flath, D. R. Black, D. G. Guadagni, W. H. McFadden, and T. H. Schultz, J. Agr. Food Chem., 15, 29 (1967).

21.   W. E. Harris and H. W. Habgood, Programmed Temperature Gas Chromatography, Wiley, New York, 1966.

22.   T. H. Schultz, R. R. Forrey, and T. R. Mon, J. Food Sci., 35, 165 (1970).

23.   A. Zlatkis, D. C. Fenimore, L. S. Ettre, and J. E. Purcell, J. Gas Chromatog., 3, 75 (1965).

24.   R. Teranishi, R. A. Flath, and T. R. Mon, J. Gas Chromatog., 4, 77 (1966).

25.   J. Pypker, in Gas Chromatography-Edinburgh 1960 (R. P. W. Scott, ed.), Butterworths, London, 1960, p. 240.

26.   T. R. Mon, R. A. Flath, R. R. Forrey, and R. Teranishi, J. Gas Chromatog., 5, 409 (1967).

27.   D. W. Carle and T. Johns, ISA Proc. 1958 Natl. Symp. Instrumental Methods of Analysis, Houston, Texas, 1958

28.   T. R. Mon, USDA, Albany, Calif., unpublished results.

29.   R. Teranishi, R. E. Lundin, W. H. McFadden, T. R. Mon, T. H. Schultz, K. L. Stevens, and T. Wasserman, J. Agr. Food Chem., 14, 477 (1966).

30.   W. S. Ryder, in Flavor Chemistry, Advances in Chemistry Series 56 (R. F. Gould, ed.), American Chemical Society, Washington, D.C., 1966, p. 70.

31.   G. H. Fuller, R. Steltenkamp, and G. A. Tisserand, in Recent Advances in Odor: Theory, Measurement, and Control (H. E. Whipple, ed.), Ann. N.Y. Acad. of Sci., 116, Art. 2, 711 (1964).

32.   H. S. Knight, Anal. Chem., 30, 2030 (1958).

33.   Hamilton Co., Whittier, Calif.

Chapter 5

MASS SPECTROMETRY: INSTRUMENT REQUIREMENTS AND LIMITATIONS

I. INTRODUCTION

A. Requirements of Flavor Chemistry

Mass spectrometry has become increasingly important in the field of structure elucidation and compound characterization in organic chemistry. Many organic chemists now include mass spectral data, along with IR, NMR, and UV spectra, among the physical properties used for characterization of new synthetic compounds. The field of organic mass spectrometry has developed rapidly, with major development occurring during the last 15 years. This rapid growth has been dependent to a great extent upon the availability of commercial instrumentation appropriate to the requirements of organic chemists. Flavor chemists were among the first to recognize the applicability of mass spectrometry to their problems of structure determination. As early as 1951, Hercus and Morrison (1) employed mass spectrometry for identification of volatile compounds produced by apples. By 1957, mass spectrometry was established as an appropriate, if not generally available, method for identification of volatile food components, primarily by investigators at what is now the U.S. Army Natick Laboratories (2). For the 10-year period between 1955 and 1965, flavor chemists recognized the importance and applicability of mass spectrometry in their work, but the expensive instrumentation was not generally available. Since 1965, mass spectrometers have become more or less generally available to flavor chemists and are now regarded as indispensable tools in the study of the chemistry of natural food flavors.

This discussion deals with the instrumentation problems unique to the requirements of chemists involved in investigation of natural food flavors. A number of texts, including those by Biemann (3), Budzikiewicz et al. (4), and McLafferty (5), concentrate on interpretation of mass spectra and the relationships between spectra and structures of organic compounds. Approximately 6000 spectra have been collected from various sources and published by Stenhagen et al. (6). Recently a monthly publication, The Journal of Organic Mass Spectrometry, has appeared and is devoted entirely to relationships between mass spectra and structures of organic compounds. The book by Beynon (7) has served as the classic general reference for mass spectrometry instrumentation. More recently, instrumentation and techniques have been treated in a book by Roboz (8). Mass spectrometry has become an important and integral part of organic chemistry and is treated to some extent in most of the recent textbooks used in introductory organic chemistry courses.

Mass spectra are not physical properties of molecules. They are dependent, in large measure, on the design of the instrument used for recording the spectra and the manner in which it is operated. Therefore, it is important to consider those operating parameters and conditions which influence the mass spectral pattern obtained. Our attention will be directed not to principles of design of mass spectrometers, but to those design features which significantly influence the quality and type of data produced and which determine the sensitivity, that is, the minimum amount of sample required for recording useful data. It is this latter requirement which is often of most significance in studies of natural flavors. Amounts of materials available are far smaller than those generally employed in other fields of organic mass spectrometry, and one is usually faced with the problem of how to extract the maximum amount of information from the minimum amount of sample.

The requirements of the flavor chemist for methods of characterization of organic compounds are more stringent than those required in most other fields of organic structure determination. Generally, the amounts of sample are very small, in the microgram or submicrogram range. The dynamic range of sample amounts is extremely large; a single mixture may contain milligrams of some components and nanograms of others. The picture is further complicated by the fact that the quantitatively minor components may have the most flavor significance or may have a subtle but important influence upon the character of the flavor. Generally, the significant contributors to natural food flavors are present in concentrations on the order of fractions of parts per million to a few parts per million of the total mass of food. The situation is somewhat more favorable in studies of essential oils. In the latter, a preliminary concentration step, usually distillation or extraction, has been carried out before identification of the components is necessary. The flavor chemist, however, usually starts with a food sample and must isolate, separate, and identify its important flavor components.

The flavor chemist's job is, in some sense, similar to those of the biochemist, pharmacologist, and toxicologist: he is interested in small quantities of organic chemicals present in very complex mixtures, the components of which exert some physiological effect. In flavor research this physiological effect is contribution to flavor. Mass spectrometry has become an indispensible tool to flavor chemists. It is, perhaps, the most useful method for identification and structural characterization of organic compounds, even though differentiation between some isomers, particularly aliphatic double-bond position, epimers, and other closely related structures is often not possible. Mass spectrometry has one additional limitation. It is, in many cases, the most sensitive analytical method available, but its sensitivity or selectivity is not comparable to that of the human nose, the ultimate instrument in flavor chemistry.

B. Information Provided by Mass Spectra

A number of factors are involved in the recent rapid growth in the use of mass spectrometry in all fields of organic chemistry. It is one of the most sensitive physical tools available to the chemist, though few workers require or use all of the sensitivity available. (Also, sensitivity varies over very wide limits depending on instrument operating parameters and method of sample introduction. The factors which influence sensitivity will be discussed in Section VII.) In addition to high sensitivity of mass spectrometers, the relatively simple correlation between spectra and structure is significant in the general acceptance of the method by organic chemists. The reactions which occur in a mass spectrometer ion source usually can be predicted on the basis of structure and reactivity under other conditions. The highly specific structural data and high sensitivity provide correspondingly high information content per microgram of sample. This can be extended to a more general statement, that mass spectrometry provides the highest structural information content per microgram of sample per second of effort expended by the chemist. Modern commercial mass spectrometers afford rapid scanning and recording of complete spectra in a few seconds. Interpretation may require significantly more time. The sample requirements of modern mass spectrometers are compatible with high efficiency separation methods, for example, capillary column gas chromatography and thin layer chromatography as well as microscale column chromatography. It is usually possible to get some structural data from a mass spectrum of any material that is detectable by gas chromatographic ionization detectors. Since the mass spectrometer is sensitive to all compounds present in the ion source, there is an overriding requirement for high sample purity; therefore, the highest separation efficiency is required prior to mass spectrometric examination of compounds.

One of the most important pieces of information which can be observed in mass spectra is the mass of the molecular ion, when it is present. This immediately gives the molecular weight of the unknown compound. Unfortunately, not all organic compounds exhibit significant molecular ion intensities in their mass spectra under the usual conditions of ionization by electron impact. Fragment ions are produced after the molecular ion is formed, and much detailed structural information can be determined by interpretation of the fragmentation pattern of the molecular ion. Peaks from stable isotopes appear in mass spectra; and in many cases, when a molecular ion is present, the isotope peaks provide information concerning the elemental composition of the molecule. Diffuse peaks (metastables) are also observed in many spectra and these provide information relating to the decomposition mechanism of the molecular ion and the fragment ions. Metastables are becoming increasingly important as sources of structural information. High resolution mass spectrometers, instruments capable of resolving ions of the same nominal mass but different elemental composition, have become commercially available. The additional structural data available from high resolution spectra have aided in interpretation of decomposition mechanisms and in resolving ambiguities in interpretation of spectra.

C. General Characteristics of Mass Spectrometers

Mass spectrometers differ in dimensions, manner of ion production, and type of operation but they all have a number of common features. It is necessary to maintain a high vacuum, at all times, in most parts of the mass spectrometer. Operationally, this imposes some serious restrictions on sample manipulation. Some kind of specialized inlet system is required to transfer the sample from the atmospheric environment to the ion source of the mass spectrometer. The ion source is perhaps the most critical part of the instrument. It is here that most of the kinds of ions described above are produced, and it is in the ion source that the nature of the spectral pattern is,

for the most part, determined. Ions produced in the ion source are
separated in the analyzer portion of the instrument. This may be an
electromagnetic separation system or an analyzer based upon other
means of separating charged particles. All instruments have ion collectors and these vary in their efficiency. They are the detection part
of the instrument. Attached to the collector is an ion current amplifier, the output of which is conditioned by a signal processing system.
Output of the signal conditioning system is recorded either on paper or
magnetic tape or is processed directly by a computer.

D. Resolving Power

Before further discussing instrumentation for mass spectrometry,
it is necessary to define resolving power and to provide some qualitative definitions of resolving power ranges. Figure 5-1 illustrates two
partially resolved mass spectrometric peaks. Definitions of resolving
power are usually based on the mass at which the valley, that is, the
minimum point between two peaks, is some fixed proportion, 1-10%,
of the peak height. Another frequently encountered definition of resolving power is based upon peak "cross-talk" or contribution of adja-
cent peaks to peak height of a specific ion peak. The 1% peak height contribution is sometimes employed for specifying instrument resolving
power, but the 10% valley definition is more common. The relationships between these definitions of resolving power are indicated in
Fig. 5-1.

Mass spectrometers are usually divided into three classes,
depending on their resolving power capability. Low resolution instruments, for residual gas analysis in high vacuum systems, have resolving powers up to about 150. These will not be discussed further.
Medium resolution instruments, those most commonly used for
analytical work and in organic structure determination, have resolving
powers between 500 and 10,000. High resolution instruments are
generally considered to be those with resolving powers higher than

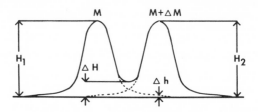

FIG. 5-1   Definitions of resolving power (RP). $H_1 = H_2$.
RP = M/ΔM.   % Valley = 100 ΔH/H_1.
% "Cross-talk" (interference) = 100Δh/H_1.
$RP_{(50\% \text{ valley})} \approx 1.3 RP_{(10\% \text{ valley})}.$
$RP_{(1\% \text{ interference})} \approx 1.9 RP_{(10\% \text{ valley})}.$

10,000. Modern high resolution instruments attain resolving powers
greater than 100,000. For purposes of this discussion, we will define
only two resolving power ranges. We will consider low resolution
spectra to be those in which the resolving power is equal to the highest
mass encountered in the spectrum (this will include most of the range
of the medium resolution instruments defined above); high resolution
spectra, those in which mass doublets differing by less than one part
in 10,000 are resolved.

II. INLET SYSTEMS

A. Gases and Volatile Liquids

The function of the inlet system is to introduce the sample,
generally at atmospheric pressure, into the mass spectrometer ion
source where pressure must be maintained below 10^{-5} Torr. A
variety of inlet systems have been described and these have been
reviewed in the standard texts (3, 7, 8) on mass spectrometry instrumentation. Only a few examples will be described here. Figure 5-2 shows an inlet system for gaseous samples. A gaseous

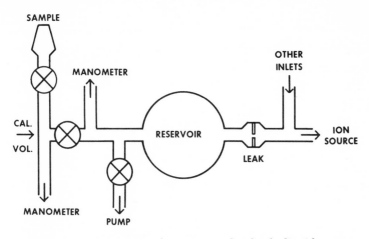

FIG. 5-2 Inlet system for gases and volatile liquids.

sample typically would be contained in a glass bulb equipped with a
stopcock and standard taper connection by which the bulb is attached
to the inlet system. The valves in this system could be glass stopcocks; if lubricating materials must be avoided, Teflon stopcocks may
be employed. Gas samples are admitted to the calibrated volume
(usually 1-3 cc) and the sample container is closed. After the inlet
system has been thoroughly evacuated, it is isolated from the vacuum
pump and sample is allowed to expand into the reservoir. The sample
pressure may be measured either before or after expansion of the
sample. For quantitative analysis of mixtures, precision micro-
manometers are employed, but these are not generally necessary for
organic structural work. Pressure in the sample reservoir is usually
in the range of 10^{-3} to 10^{-1} Torr. The leak, usually a gold foil with
a small hole or a sintered glass or metal disk, is selected so that the
sample depletion rate is small (2-3%/hr) during the time required for
recording the mass spectrum. If stainless steel valves are substituted for stopcocks and the entire system is enclosed in an oven, this
type of inlet may be heated to over 150° C and used for analysis of
volatile liquids as well as for gases. Use of magnetically activated
all-glass valves extends this range to over 300° C.

For analysis of sample quantities in the range of 10-100 µg, a similar inlet system may be used with a smaller reservoir, or the reservoir
may be bypassed entirely. Further enhancement of sensitivity may be
attained by eliminating the leak. When sensitivity is enhanced in this
manner, sample depletion rates are very rapid and the sample may be
consumed in a few seconds, requiring a short spectrum recording
time. Ryhage (9) described a heated inlet system, which can maintain a temperature up to 350° C, that provides a leak and reservoir
which may be employed or bypassed and is appropriate for introduction
of samples in the microgram range.

A diagram of a simple inlet system for small samples is shown in
Fig. 5-3. This system was described by McFadden (10), but variations
of this sample introduction technique have been used in many laboratories. It is particularly appropriate for analysis of very small
quantities of materials collected from gas chromatographic columns
and thin layer chromatographic plates when a fast scanning mass
spectrometer is available. The sample is placed in the tube, D, and
attached by a vacuum connection, C, to the inlet system. The
micrometer valve, E, is closed and the system is evacuated by opening
valve B. During this initial evacuation, the tube can be cooled in
liquid nitrogen to prevent vaporization of the sample. When air has
been removed, vacuum valve B is closed, isolating the inlet from the

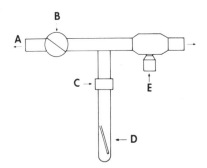

FIG. 5-3   A simple, versatile mass spectrometer inlet. A,
            auxiliary vacuum, B, vacuum valve, C, quick-fit
            connector, D, sample, E, micrometer valve (10).

vacuum system. The micrometer valve, E, is opened slowly while the
sample tube is warmed carefully, either by lowering the level of
coolant or by applying external heat when a sample of relatively low
volatility is being examined. Sample temperature is increased until
adequate partial pressure of sample in the ion source is attained and
the spectrum is recorded. If the sample size is very small, the
micrometer valve may be opened further to allow a greater flow rate
into the mass spectrometer. With careful control of temperature and
manipulation of the vacuum valve, solvents may be removed in this
inlet system. Samples containing a mixture of components of wide
boiling range may be fractionated on such an inlet by programming
the temperature either continuously or stepwise. Such a technique
was used by Merritt et al. (11) for identification of the volatiles isolated by vacuum distillation of coffee.

B. Liquids and Solids of Low Volatility

Liquids or solids of low vapor pressure may be introduced directly
into the mass spectrometer ion source. When a direct introduction
system is used, thermal decomposition may be significantly reduced
since the sample partial pressure required is of the order of 10^{-7} to
10^{-6} Torr at source temperature (usually 250° C). Vapor pressures
of this order are readily attained even with materials not ordinarily
considered to be volatile, such as free amino acids (12) and derivatives
of polypeptides (13,14). A direct inlet system requires a vacuum lock
and usually a special pumping system in order to place the sample as
close as possible to the ionizing electron beam, usually less than 1 cm
from the beam.

Direct inlet systems are available for most commercial mass
spectrometers and are almost indispensable if maximum versatility
is required of the instrument.

C. Collected GC and TLC Fractions

A direct introduction system was used by Amy et al. (15) for recording high resolution mass spectra of sample quantities on the

order of $0.1\ \mu g$, collected from a gas chromatographic column effluent. The collection tubes each contained a few milligrams of column
packing to increase collection efficiency, which is predictable on the
basis of chromatographic behavior. The sample can be introduced
into the mass spectrometer ion source with the direct introduction
system without removing the sample from the column packing material,
if a low volatility packing is employed.

Even if the direct introduction system and ion source are not
heated, or are operated near room temperature, there are obvious
upper limits on volatility of samples which can be introduced in this
manner. Volatile samples may be lost during evacuation of the inlet
system, prior to insertion of the sample into the ion source. Damico
et al. (16) introduced samples as volatile as propionaldehyde via the
direct inlet system after collection in tubes packed with a few milligrams of activated charcoal. Adsorption on charcoal reduced volatility
sufficiently so that the samples could be introduced without loss,
though this was not possible when ordinary chromatographic column
packing was used in the collection tubes.

Direct introduction systems are also very useful for examination
of samples separated by thin layer chromatography, provided volatility
is low enough. A sample can be eluted from the silica gel with a
solvent and transferred to the inlet system sample container. Solvent
can then be evaporated before or during evacuation of the direct inlet
system.

D. Gas Chromatographic Inlet System

One of the most widely used and most versatile mass spectrometric
sample introduction systems is the combined gas chromatograph-
mass spectrometer. The chromatograph is connected directly to the
mass spectrometer through an enrichment device or sample splitter,
and the separation power of the chromatograph can simplify the problems of mass spectrometric identification of very small samples.
This topic will be treated in Chapter 6.

E. Inlet Systems Required in Flavor Investigations

Emphasis must be placed on versatility in selection of sample inlet systems. The flavor chemist is often faced with samples that do not meet the conventional requirements of gas, liquid, or solid inlet systems. Many liquids are more appropriately handled in the vapor phase or in an inlet system designed for handling solids. A liquid introduction system that depends on a syringe and septum for admitting samples is useless if there is not enough liquid to fill the syringe. In this case, one of the simple introduction systems described above may be most appropriate. The gas chromatographic inlet and direct inlet systems are probably the most versatile available; however, a liquid introduction system should be provided. It is useful for introduction of mass marking reference standards during analysis of samples introduced through the other inlet systems.

III. ION SOURCES

A. Electron Impact

The electron impact ion source is the most widely used in mass spectrometry. Ions are formed by collision with an electron beam of adjustable energy in the range of 0-100 eV. Electron energy of 70 eV is most common. The general reaction involved in formation of a positively charged molecular ion is:

$$M + e^- = M^+ + 2e^- \tag{1}$$

where M represents the original molecule, and M^+ is the molecular ion. When 70-eV electrons are used, the molecular ion is formed with excess kinetic energy and decomposes, in a manner depending on its structure, to yield a variety of fragment ions which provide the characteristic mass spectral pattern. Energy of the bombarding electron beam significantly influences kinetic energy of the resultant ions. The relative intensity of the molecular ion peak depends very

strongly on electron energy. At energies close to the ionization potential of the molecule, the spectrum is seen to consist almost entirely of
the molecular ion for compounds in which the molecular ion is relatively stable. As electron energy is increased from the threshold of
ion production, usually 9-12 eV, a greater number and variety of fragment ions are formed. The molecular ion and its stable isotopes are
very important for determining the molecular weight of an unknown
compound. The fragmentation pattern is important in determining
structure of the unknown material, and many empirical correlations
based on known chemistry and reactivity of a large variety of organic
compounds have been reported.

The ion source is the reaction vessel in which fragmentation
occurs. The pattern is determined primarily by reaction conditions
in the source. An electron impact ion source is shown schematically
in Fig. 5-4. Many variations are employed in the exact geometrical
relationships found in commercial instruments, but that shown in
Fig. 5-4 is typical of those used with magnetic deflection mass
spectrometers. Source pressure may be maintained on the order of
10^{-7} Torr by the main analyzer vacuum system. For maximum
versatility, higher source pressures are desirable. Differential
pumping, by a separate vacuum system, permits operation of the
source at pressures in the range of 10^{-5} to 10^{-4} Torr. This range is
too high for satisfactory analyzer operation. Either the source exit
slit S_3 or a pumping baffle serves to isolate the source vacuum system
from the main analyzer. Sample enters through the gas inlet tube,
usually heated to about 200° C, and the components of the sample are
ionized by electron impact in the electron beam, shown in the diagram
as a crosshatched area.

Electrons are produced by thermionic emission from the filament,
F, a tungsten or rhenium wire heated by an electrical current to
at least 2200°K. Usually a 3- to 4-A heater current is employed.
The heater should be provided with direct current to avoid modulation
of 60-cycle noise on the electron beam and resultant ion beam.

Kinetic energy of the electron beam is determined by the potential difference between the filament and the slit, S_1. This energy can be controlled on most instruments in the range of 0-100 eV. A relatively low intensity (100-300 G) magnetic field is applied with magnets mounted either external to the ion source or inside the source itself to collimate the electron beam. The collimating magnetic field is parallel to the electron beam shown in Fig. 5-4. Most instruments have provision for measuring total filament emission current and electron current arriving at the anode (target or trap). Circuitry for automatically controlling total emission or anode current is generally provided.

The electron impact ion source is a relatively inefficient device for ion formation. Ionization efficiency depends, in part, upon the nature of the molecular species involved, but for electron currents of $100~\mu A$ the maximum ion current which can be produced at a partial pressure of 10^{-6} Torr is of the order of 10^{-7} A (8). Space charge

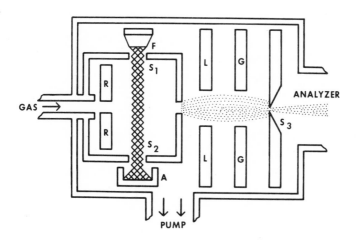

FIG. 5-4   Electron impact ion source. R, repeller, F, filament, A, anode, S_1 and S_2, electron beam defining slits, L, lens or focusing electrodes, G, accelerating electrode, S_3, source exit slit.

effects and inefficiency of ion extraction and other losses result in the
availability of only a small fraction of this ion current for analysis
after removal from the ion source. Generally, ion currents in the
range of 10^{-8} to 10^{-11} A are produced. The repellers, R, are provided with a small positive voltage relative to the ionization chamber
(1-15 V) to aid in the extraction of ions. The lens electrodes, L,
focus and center the beam on the accelerating electrode, G. Usually,
the entire ionization chamber is maintained at a high positive potential
which may vary from 2 to 10 keV. The source exit slit, S_3, which
may be adjustable, determines resolving power and ion current. In
addition to aiding in the ion extraction process, the voltage applied to
the repellers determines the residence time of ions in the ionization
chamber. Higher repeller potentials result in more rapid extraction
or reduced residence time. Thus, repeller potential can significantly
influence the mass spectral pattern, particularly those peaks resulting
from relatively slow reactions. Repeller potential has significant
effect on the intensity of metastable peaks in the mass spectra.

Figure 5-5 is a schematic diagram of the ion source of a time-of-
flight mass spectrometer. The electron gun and gas inlet system are
almost identical to those of the source shown in Fig. 5-4. The major
difference is in the mode of extraction of ions. The electron beam may
be either continuous or pulsed, but ion extraction is always pulsed
in the time-of-flight instrument. The ionization chamber is maintained
at ground potential. Negative extraction pulses of about 250-V ampli-
tude and duration of 1-3 μsec are applied to grid G_1. These pulses
are applied at repetition rates that may be varied between 2000 and
50,000/sec. Grid G_2 is maintained at a negative dc potential of about
3000 V. Ions are extracted from the source and enter the flight tube,
in short bursts, with kinetic energy of about 3 keV. No slits are employed and the transmission of the grids is about 90%. Thus, with
two grids, overall transmission of the ion beam is about 80%. Ionization efficiency is of the same order of magnitude as that described for
the source shown in Fig. 5-4. The time-of-flight ion source may be

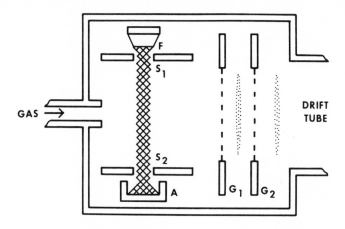

FIG. 5-5   Time-of-flight mass spectrometer ion source. F,
           filament, A, anode, S_1 and S_2, electron beam de-
           fining slits, G_1, ion extraction grid, G_2, accelerat-
           ing grid.

operated in a pulsed mode, that is, with the electron beam turned on
for a very short interval (less than 1 μsec) during the instrument
cycle, which may be as long as 100 μsec. In the continuous ionization
mode, the electron beam is "on" continuously and ions are maintained
in a "potential well" between accelerating pulses. This mode of operation results in greatly enhanced sensitivity of the time-of-flight instrument but requires very careful balance of dc bias potentials and pulse
shape and amplitudes (17).

The usual ion source operating temperature is maintained at about
250°C but may be reduced for analysis of samples of low thermal
stability. Minimum temperature is fixed by the heat produced by the
hot filament. The ion source of the time-of-flight instrument is
usually operated at room temperature.

The ion source shown in Fig. 5-4 is provided with a vacuum system separate from that of the main analyzer vacuum system. Such an
arrangement is highly desirable in that it provides for maintaining
source pressures higher than those normally permissible in the rest

of the instrument. Differential pumping is particularly desirable when a gas chromatographic inlet system is used or when a large proportion of the samples are introduced via the direct introduction system and analyzer contamination must be maintained at a minimum. A source isolation valve, which permits removal of the ion source without breaking the analyzer vacuum, is highly desirable in a commercial instrument but is available in few. In an instrument that receives heavy use, it is necessary to remove, dismantle, and clean the ion source periodically. Filaments must be replaced at intervals varying between 2 and 6 months, depending on how often the mass spectrometer is used and the types of samples to which it is exposed.

Attempts have been made to increase the ionization efficiency and overall efficiency of ion recovery from the electron impact ion source. An ion source has been described (18) in which the ion beam is produced by an electron beam that is coaxial with the molecular sample beam passing through the ion source. This increases contact time between the electron beam and the gas sample and therefore increases ionization efficiency. Another approach involves the use of electrostatic potentials applied to focus the ion beam on the exit slit without the losses that usually occur when the ions pass through this slit (19). Improvements of sensitivity of the order of 30-100 times have been reported with electrostatic focusing of the ion beam.

Most commercial mass spectrometers provide a total ion current monitor electrode. In single-focusing magnetic instruments it is usually located in the source or between the source and the analyzer. The monitor electrode may be one of the existing functional source electrodes (for example, the exit slit may be connected to an electrometer amplifier to indicate total ion current) or an additional electrode may be placed in the path of the ion beam emerging from the source to intercept a fixed percentage (usually between 10 and 50%) of the total ion beam (20).

B.   Field Ionization

Other sources of positive ions, fundamentally different from the electron impact ion source, have been described and may become significant in their applications in flavor chemistry. The field ioniza-tion technique has been applied to the study of lipids (21) and pesticides (22). In a typical field ionization source, an anode emitter maintained at a positive potential between 2 and 6 keV extracts electrons from molecules by the high electric field in the immediate emitter vicinity. Typical fields of 10^7-10^8 V/cm are obtained with emitters of 100- to 500-Å radius of curvature at voltage differences between emitter and cathode of 500-25,000 V. Total ion currents are generally less than 10^{-8} A, and transmissions are of the order of about 1%. Generally, electron multipliers or photoplates are used as detection devices. Various emitter geometries have been evaluated and different types are preferred by different investigators. Performance of sharp metal edges and thin wires as field emitters was compared by Beckey et al. (23). They concluded that wires and sharp edges gave higher sensitiv-ity and better stability than the metal tips originally used in field ion sources. A combined field ionization-electron impact source, which can be operated in either mode, was described by these workers. A comparison of spectra obtained in the two different modes is shown in Fig. 5-6. The major difference between field ionization and electron impact spectra is apparent in the spectra of n-heptane. In the field ionization spectrum, fragment peaks are essentially absent and almost all of the total ionization is accounted for by the molecular ion. In general, most field ionization spectra exhibit some fragment ions as well as metastable peaks, but the spectra are usually much simpler than those observed from electron impact sources. Because of the relatively high proportion of molecular ions produced in the field ionization source, it is likely that this technique will be more widely applied in the near future and will be a valuable one for chemists.

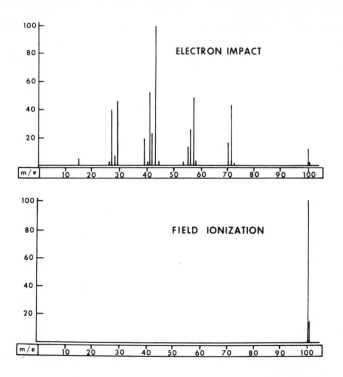

FIG. 5-6   Mass spectra of n-heptane obtained by means of a
            combined electron impact-field ionization ion
            source (23). (Reproduced with permission of the
            Institute of Petroleum.)

Recent developments in field ionization mass spectrometry have been
reviewed by Block (24).

C. Chemical Ionization

Chemical ionization mass spectrometry has been reviewed by
Field (25). This method employs a basically different mechanism, in
which ionization is brought about by collision of molecules with charged
ionic species. Most work in chemical ionization mass spectrometry
has used methane as the source of ions. Generally, some modification
of the ion source is required so that partial pressure of methane in the
source can be produced and maintained at about 1 Torr. This

requires very efficient differential pumping so that pressure in the analyzer can be maintained at the level of 2×10^{-5} Torr. Since the reactions producing the ions are fundamentally different from those produced by other ionization methods, the spectra potentially can yield information about structure of unknown compounds quite different from that available in conventional mass spectra. The technique will undoubtedly become more useful in the near future as more investigations are made of the behavior of different types of compounds under the conditions in a chemical ionization ion source.

D. Photoionization

Production of ions by impact with photons has been employed in fundamental chemical physics investigations (26). Generally, rather elaborate instrumentation is required for construction of photoionization sources for mass spectrometry. This technique also can be potentially useful in structural studies because of the ability to carefully control the energy of the photons which bombard the unknown molecule. Ions with extremely small excess kinetic energy are produced, resulting in intensification of the molecular ion peaks in mass spectra of compounds which produce relatively low intensity molecular ions under electron impact.

IV. MASS ANALYZERS

The principles, design, and details of construction of mass analyzers for mass spectrometers have been described by Beynon (7), Roboz (8) and others. Here, we discuss only those factors of most importance in utilizing the instruments in flavor chemistry and only those instruments most commonly employed in structural studies, namely, magnetic deflection, time-of-flight, and quadrupole mass analyzers.

A. Magnetic Sector Analyzer

In the magnetic deflection mass analyzer, the ions produced in any of the sources described above and accelerated to relatively high velocities by a potential difference, V, leave the ion source with kinetic energy $eV = 1/2 mv^2$, where m is the mass of the particle, e is the electronic charge, and v is the velocity. If this beam passes through a magnetic field oriented perpendicular to the direction of flight, the beam is curved by an accelerating force, Hev, where H is the magnetic field strength. To maintain a stable path, the accelerating force is equal to the radial force, $Hev = mv^2/r$, where r is the radius of curvature of the ion path. Elimination of the velocity term leads to the familiar equation describing the flight path of ions in the mass spectrometer:

$$\frac{m}{e} = \frac{H^2 r^2}{2V} \qquad (2)$$

Radius of curvature is fixed by the geometry of the particular mass analyzer used. Ions of different mass-to-charge ratios are brought to focus at a collector slit by variation of magnetic field or accelerating voltage.

In the elimination of the velocity term, according to the classical derivation of the equation relating m/e to magnetic field strength, radius of curvature, and accelerating voltage, it is assumed that v is constant. Actually, a conventional electron impact ion source produces an ion beam with a kinetic energy distribution. The energy distribution limits the resolving power attainable with a single magnetic sector. This device is not strictly a mass spectrometer but is rather an energy or momentum spectrometer.

Commercial instruments are available with radii of curvature in the range of 5-12 in. Deflection angles normally are 60, 90, or 180°. In general, the total pathlength traversed by the ion beam increases as the deflection angle decreases in a fixed radius instrument; therefore, the path is shorter at greater deflection angles. Short pathlengths are likely to result in smaller losses due to collision and scattering by molecules in the path of the ion beam. Pathlength may also have a significant influence on the spectral pattern observed, particularly with respect to fragment ions formed at a relatively low rate during flight through the analyzer. As flight time increases, intensity of the parent peak decreases and fragment peak intensities increase. Generally, peaks representing product ions from metastable transitions exhibit this effect. A similar effect of accelerating voltage has been observed and reported (27). Pathlength and accelerating voltage are just two of the factors which may be responsible in part for differences in mass spectra consistently observed between instruments of different geometries. These are, however, systematic differences which can be corrected or, at least, reduced in instruments of the same dimensions and design.

Resolving power of the magnetic sector mass analyzer is determined by the exit slit width setting and by the corresponding dimension of the collector slit. Resolution is also influenced by accelerating voltage. Since most of the ion sources discussed above produce an ion beam with a relatively small distribution of energies, this energy distribution is the limiting factor, along with various aberrations, in determining the resolving power of a single-focusing magnetic deflection instrument. The relative effect of the energy distribution becomes more significant at lower accelerating voltages, so in principle, at least, the highest possible accelerating voltage should result in highest resolving power. Resolving powers on the order of 10,000 have been reported for single-focusing magnetic sector instruments with radii of curvature in the 8- to 12-in. range (28-30). These values represent the ultimate attainable under optimum conditions and generally cannot be expected in routine use where ion sources may be dirty or electrical instabilities may be present in the circuitry. In general, however, routine resolving power of about 3000 can be attained in these single-focusing magnetic deflection instruments. These resolving powers have been shown to be adequate for precise mass measurement with an accuracy of 10 ppm (31), a range adequate for unambiguous

determination of the elemental formulas of many organic ion peaks
appearing in spectra of relatively low molecular weight compounds.
Ion transmission and, therefore, instrument sensitivity are decreased
by reducing the slit widths to those required to attain these high
resolving powers, and, thus, maximum sensitivity is obtained at minimum resolving power settings. Since the flavor chemist usually
requires maximum sensitivity for identification of the extremely small
quantities available to him, it is most practical to employ the minimum
resolving power necessary to solve a specific problem.

Spectra may be produced in magnetic analyzers by scanning either
the magnetic field or accelerating voltage. Variation of accelerating
voltage has become less popular in recent years, since at low
accelerating voltages sensitivity of the instrument is diminished, resulting in discrimination against higher masses. Magnetic scanning
instruments are becoming more popular, but these suffer the disadvantage of a nonlinear mass scale and difficulties of reproducing the
mass scale on a strip chart or other recording device due to hysteresis
in the magnet. This problem is more severe at the fast scan rates
used with sample introduction by direct probe and by the gas chromatographic inlet system.

The specific function chosen as the relationship between time and
mass has a significant effect on the measured mass spectral pattern.
It is usually assumed that peak heights are proportional to the number
of ions. If peak widths are not equal for all masses, this assumption
is not correct. Use of a decreasing exponential scan, i.e.,

$$\frac{m}{e} = \frac{M_o}{e} \exp(-t/T) \tag{3}$$

where M_o is the starting mass, t is time, and T is the time constant
of the scan, provides equal peak widths. Therefore, peak height is a
correct measure of the number of ions only if an exponential scan is
used in the magnetic deflection spectrometer. Other scan functions
will lead to nonuniform peak widths and result in some difficulty in

comparing spectra recorded on instruments employing different scan
functions. These differences are systematic and can be corrected
during computation of relative intensities if the scan function used is
known.

B.   Time-of-Flight Analyzer

The time-of-flight mass analyzer was one of the first analyzers
used for rapid analysis of gas chromatographic effluents without the
need for preliminary sample trapping (32). This instrument is inher-
ently simple, since the analyzer consists of a long (about 180 cm) drift
tube maintained at the accelerating voltage employed in the ion source.
The narrow pulses of ions ejected from the ion source enter the drift
tube with the same kinetic energy, and, therefore, if a long enough
drift length is provided they will separate according to mass because
of the different terminal velocities achieved by different masses at
this constant energy. Ions with small m/e values arrive at the end of
the drift tube a few microseconds after leaving the source, while ions
of m/e 200 or higher may require 20-30 μsec to traverse the same
distance. Complete spectra are produced at the rate of 10,000-50,000/
sec and can be observed at this speed only through use of an oscilloscope synchronized with the ion source pulse circuitry. These high
repetition rates provide an essentially continuous display of the spectrum on the oscilloscope screen, which is very convenient for the
operator. The condition of the instrument and its operation can be
monitored continuously, even while spectra are being recorded via
alternate output systems, on an oscillograph recorder or other data
processing systems. This continuous oscilloscope monitoring is one
of the primary advantages of the time-of-flight instrument. Its useful
recording speed with the analog output gating system provided by the
manufacturer is comparable to that attainable in modern magnetic
deflection instruments (1-30 sec), but the absence of the magnet in
this system is a distinct advantage. The major disadvantages of the
time-of-flight instrument are the requirement for high frequency

electronic circuitry and the limitation on its attainable resolving power (usually a maximum of 500). The time-of-flight mass spectrometer was the first of the commercial instruments available with an electron multiplier (33). It required a wide band electron multiplier in order to detect the extremely sharp ion peaks produced (peak widths are only a few nanoseconds in duration). The multiplier was essential to provide the necessary response speed.

C. Quadrupole Analyzer

A diagram of the quadrupole mass analyzer is shown in Fig. 5-7. The ion source employed is similar to the electron impact sources used with magnetic deflection instruments. It shares an advantage with the time-of-flight analyzer, namely, no exit slit is required and the loss in ion beam intensity inherent in the use of an exit slit is eliminated. Ions are extracted from the source and focused by electrostatic lenses into a beam that enters the quadrupole field. Precisely machined and aligned cylindrical metal electrodes are connected so that a hyperbolic electric field is produced. Spatially opposite rods of the quadrupole assembly are electrically connected together, creating two rod pairs. A radio frequency signal (in the range of several megahertz) and a dc bias potential are applied to each pair in such a manner that one pair is 180° out of phase with the other pair and the dc bias of one pair is of opposite polarity to that of the other. This arrangement of electrodes and biases produces a "filter" which will permit transmission of a very narrow mass range for a given setting of the fields. A spectrum is produced by varying the dc bias so that ions of successively increasing mass-to-charge ratio pass through the filter and arrive at the electron multiplier. The entire assembly is quite compact, the mass filter required for most analytical applications being only about 6 in. long. Repetitive scanning of mass is attained by applying the dc bias potential in the form of a ramp function.

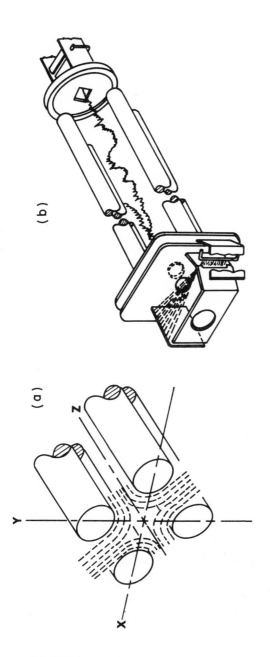

FIG. 5-7   Quadrupole analyzer. (a) Quadrupole rod structure with the hyperbolic field lines, (b) mass spectrometer filtering. (Courtesy of
Finnigan Instruments Corp.)

The spectrum can be displayed on an oscilloscope screen at repetition rates corresponding to scan times between 0.01 and 1000 sec for the selected mass range. When a linear voltage ramp is applied to the electrodes, the mass scale as displayed on the oscilloscope or on the strip chart recorder is linear with equal peak widths at all masses. This arrangement makes determination of mass very simple and is particularly adaptable to computer-controlled data acquisition. Resolving powers as high as 1000 at mass 500 (based on a 50% valley definition) are available on commercial instruments.

The quadrupole analyzer is compact and simple, requires no magnet, and is relatively inexpensive. For these reasons it is one of the more promising new instruments for general analytical use. Its fast scanning capabilities make it particularly appropriate for use with a gas chromatographic inlet system.

The total ion current monitor on most quadrupole instruments consists of a system for integrating all of the peaks from repetitive mass scans. As in the case of most magnetic deflection instruments, an interruption in total ion current recording occurs when the mass spectrum is recorded on a chart. The time-of-flight analyzer is the only instrument in which total ion current can be recorded essentially without interruption even though other output systems are being used concurrently.

D. Double-Focusing Instruments

High resolution mass spectrometry has become increasingly important in recent years for elucidation of structures of organic compounds. Since monoisotopic atomic weights of nuclides are not exact whole numbers except for C^{12} (12.00000, by definition), accurate mass measurement can be used to distinguish between ions of the same nominal mass but differing in elemental composition. Measurement accuracy in the range of 1-10 ppm is usually required. Figure 5-8 indicates the variety of elemental compositions possible at m/e 310 for ions containing carbon, hydrogen, not more than three nitrogen atoms,

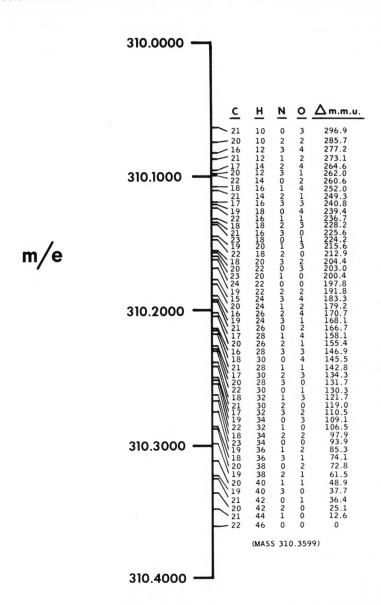

C	H	N	O	Δ m.m.u.
21	10	0	3	296.9
20	10	2	2	285.7
16	12	3	4	277.2
21	12	1	2	273.1
17	14	2	4	264.6
20	12	3	1	262.0
22	14	0	2	260.6
18	16	1	4	252.0
21	14	2	1	249.3
17	16	3	3	240.8
19	18	0	4	239.4
22	16	1	1	236.7
18	18	2	3	228.2
21	16	3	0	225.6
23	18	0	1	224.2
19	20	1	3	215.6
22	18	2	0	212.9
18	20	3	2	204.4
20	22	0	3	203.0
23	20	1	0	200.4
24	22	0	0	197.8
19	22	2	2	191.8
15	24	3	4	183.3
20	24	1	2	179.2
16	26	2	4	170.7
19	24	3	1	168.1
21	26	0	2	166.7
17	28	1	4	158.1
20	26	2	1	155.4
16	28	3	3	146.9
18	30	0	4	145.5
21	28	1	1	142.8
17	30	2	3	134.3
20	28	3	0	131.7
22	30	0	1	130.3
18	32	1	3	121.7
21	30	2	0	119.0
17	32	3	2	110.5
19	34	0	3	109.1
22	32	1	0	106.5
18	34	2	2	97.9
23	34	0	0	93.9
19	36	1	2	85.3
18	36	3	1	74.1
20	38	0	2	72.8
19	38	2	1	61.5
20	40	1	1	48.9
19	40	3	0	37.7
21	42	0	1	36.4
20	42	2	0	25.1
21	44	1	0	12.6
22	46	0	0	0

(MASS 310.3599)

FIG. 5-8   Elemental compositions possible at nominal mass of 310 (5). (Reproduced with permission of W. A. Benjamin, Inc.)

and a maximum of four oxygen atoms (5). The difference in mass (in millimass units) of each species from the mass of the saturated hydrocarbon, $C_{22}H_{46}$, is shown in the right-hand column. At higher masses, and as the number of heteroatoms increases, the possible number of species also increases. Even at m/e 100, 18 different species are possible with four or fewer nitrogen atoms and not more than four oxygen atoms (34). The availability of high resolution mass spectral data can eliminate much of the ambiguity in assignment of elemental compositions to molecular ions and fragment ions.

Basically, there are two forms of commercially available high resolution mass spectrometers, those employing the Nier-Johnson geometry and those employing the Mattauch-Herzog geometry. Both of these systems provide double focusing. They eliminate the energy distribution in the ion beam after it emerges from the ion source to provide simultaneous energy and direction focusing either at a collector slit or on a photographic plate. This results in much higher resolving powers than can be attained with single-focusing magnetic deflection instruments.

In double-focusing instruments employing the Nier-Johnson geometry, the magnetic analyzer is similar to that used in single-focusing instruments, but an electrostatic sector is employed between the source and the magnetic analyzer to provide energy focusing. With a radius of curvature of 16 in. in the electrostatic sector and 12 in. for the magnetic sector, resolving power in the range of 100,000 can be attained on commercially available instruments. With the extremely narrow slit widths required for high resolution operation, electron multiplier detection is required. Manual peak matching is a method in which the mass of an unknown peak is determined to within a few parts per million by precise measurement of the ratio of accelerating voltages required to focus the unknown peak relative to a standard reference peak (34). This system is adequate for mass measurement of a molecular ion or a few fragment peaks, but becomes very cumbersome when a large number of peaks must be measured. Comparable accur-

acy of mass measurement for small numbers of peaks can be attained
by recording multiplier output on a chart, along with a very accurate
time base signal and peaks from a reference compound. Magnetic
tape and computer systems, of course, provide better alternative solutions to the problem.

The Mattauch-Herzog geometry provides simultaneous energy and
direction focusing in a plane. The dimensions are of the same order
as those used in Nier-Johnson instruments for equivalent resolving
power. The output of the instrument can be recorded on a photographic plate placed in the focal plane, the entire spectrum being recorded simultaneously. A collector slit and electron multiplier may
also be employed. A reference compound is usually introduced along
with the sample and its spectrum is later used for comparison on a
microdensitometer equipped with a facility for precision distance
measurement along the photoplate (35). Photoplate recording provides
a compact method for recording all of the peaks (hundreds may be present) in spectra of relatively high molecular weight compounds, for
measurement and processing later, without wasting valuable instrument time.

Compact double-focusing instruments utilizing the Nier-Johnson
geometry have become available. An instrument with a 5-in. electrostatic sector and 4-in. magnetic sector is capable of providing a
resolving power of about 2500 (36). These instruments provide maximum resolving power with a relatively small magnet. They are smaller,
simpler, and less expensive than conventional single-focusing instruments, and can provide adequate performance for most flavor
investigations.

V. ION CURRENT DETECTORS

A. Faraday Cup

The simplest means of detecting the resolved ion beam is by placing
a Faraday cup or simple collector electrode behind the analyzer slit

assembly and recording the positive ion current, after amplification, on any convenient recording device. Though this system is simple, it requires extremely high impedance amplifiers and these are usually limited in their time constants. Therefore, fast scanning, which is required when very small samples are analyzed using a gas chromatographic inlet or direct introduction system, cannot be used with instruments having only Faraday cup detection systems.

B. Secondary Electron Multipliers

Electron multiplier detectors have become common on most commercially available mass spectrometers. These devices are very similar to photomultiplier tubes used in optical instruments, except that the cathode is ion sensitive rather than light sensitive. Commercial multipliers usually employ between 10 and 20 dynodes, resulting in current gain of the order of 10^6-10^7. These devices also provide much lower output impedance than is available with Faraday cups, so wider bandwidth electronics can be used for amplification of their signal output.

The high sensitivity available at the upper limit of electron multiplier gain is usually of no practical use in low resolution mass spectrometry. At this very high gain, single ions may be detected and recorded. Maximum gain is normally required only in high resolution mass spectrometry when the total number of ions available at the collector is extremely small. Figure 5-9 shows the influence of multiplier voltage on the appearance of background peaks in the mass spectrometer. The major peak appearing in the range recorded at m/e 149 arises from background produced by the diffusion pump fluid used in the inlet pumping system. It is clear that at maximum multiplier voltage (3 kV) this background can be made to go full scale on the output system, and this sensitivity is useless. The multiplier on the instrument described is usually operated at 1.2 kV, corresponding to an actual current gain of about 2000. The real advantage of the electron multiplier is that its inherent sensitivity can be used to reduce

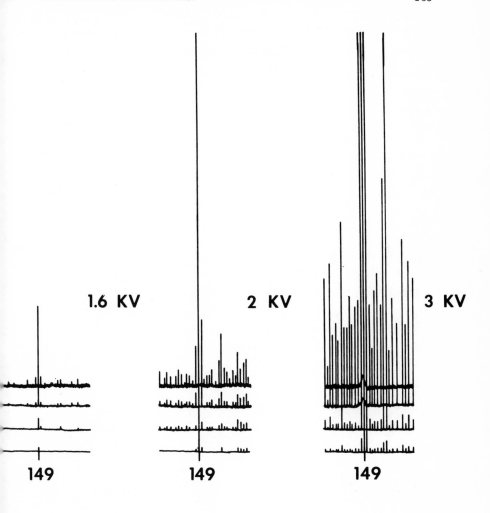

Applied Voltage (kV)	Gain
1.6	2×10^4
2	7×10^4
3	4×10^5

FIG. 5-9   Background peak at m/e 149 at various multiplier gains.

the required input impedance of the amplifier to which it is connected, thereby reducing the time constant and increasing the speed at which spectra can be recorded. In general, the sensitivity of modern spectrometers is limited not by multiplier gain but by background and noise considerations.

Ryhage (9) has discussed the statistical noise problem during fast scanning mass spectrometry employing an electron multiplier detector for analysis of sample amounts in the microgram range. His statistical calculations indicate that for a fixed time constant of the measuring system of 0.001 sec and ion currents of 10^{-11}, 10^{-12}, 10^{-13}, and 10^{-14}A, the total standard deviations are 0.4, 1.3, 4.0, and 13%, respectively. These results indicate that for reasonable precision, the ion current should not be less than 10^{-14}A. Unfortunately, the amount of sample corresponding to this ion current was not stated. Some results on the determination of attainable precision of relative intensity of mass peaks, as a function of sample consumption rate in the ion source and spectrum scanning time, are presented in Table 5-1 (37). Bandwidth of the recording system was, in all cases, 0-1000 Hz. The average relative intensity values and standard deviation estimates shown in this table were determined from six separate scans at each of the sample consumption rates and scan rates indicated. Sample was admitted via a liquid introduction system employing a reservoir to maintain constant flow rate of sample during the course of the experiment. It is clear from these data that at low sample consumption rates, variations in individual peak heights of the order at 10% on repetitive spectra recorded for the same compound at the highest recording speeds must be accepted.

The sample consumption rate of 2×10^{-9} g/sec corresponds to the rate which would be observed at the top of a gas chromatographic peak with 10-sec half-width representing 0.1 µg, passing through an inlet system with 20% efficiency; or, to a 20-nanogram sample introduced through the direct inlet and totally consumed in 10 sec. These

TABLE 5-1

Mean Relative Intensities (%) and Standard Deviation Estimates (S)
for Major Peaks in n-Amyl Butyrate Spectrum[a] (37)

Scan time, sec (m/e 20-200)	Sample consumption rate: 2×10^{-8} g/sec					
	m/e 43		m/e 70		m/e 89	
	%	S	%	S	%	S
30	83.2	2.2	55.7	3.3	66.0	1.9
3	76.0	2.7	54.5	1.2	61.0	2.0
1.5	77.8	7.9	63.4	7.2	71.0	7.6
	Sample consumption rate: 2×10^{-9} g/sec					
3	72.7	2.6	56.0	4.6	63.0	3.9

[a] Output system band width: 0-1000 Hz for all measurements.

precision measurements include noise contributions from sources
other than statistical variation in ion currents.

VI. OUTPUT SYSTEMS: SIGNAL CONDITIONING AND RECORDING

A. Amplifiers

Vacuum tube electrometer amplifiers are used in most commercial
instruments employing either Faraday cup or electron multiplier
detectors. These amplifiers provide high grid resistance and can be
used with extremely high (10^{12} Ω) load resistors employed with
Faraday cups. At these very high resistances, the circuit time constant

is quite long, on the order of seconds, because of stray capacitance in the circuitry employed. The electrometer tube may also be a source of noise that can become disturbing if the time constant is lowered by use of lower value load resistors with electron multiplier detection. This noise may be caused by microphonics in the tube. Another problem associated with electrometer tubes is their slow recovery from overloading. This can be a severe problem in mass spectrometry since very often small peaks follow very large peaks and fast amplifier recovery is required. Solid state amplifiers using field effect transistor input stages are now available. These devices provide significantly lower noise levels and lower drift rates than are attainable in vacuum tube amplifiers. They are more compact and can be mounted directly in contact with, or in close proximity to, the electrom multiplier. They are simple and inexpensive and provde a wide dynamic range, low noise, and wide bandwidth.

B. Recorders

A variety of recording devices has been used in mass spectrometry. Conventional strip chart servopotentiometers can be used for spectra that are scanned at relatively low rates. The limiting problem with a servopotentiometer is its slow response speed of about 0.2 sec. They do provide high accuracy and precision, of the order of 0.1%, and are therefore most applicable in quantitative analysis or in precision isotope ratio determination, where time is not a limiting factor. These recorders find use in most mass spectrometry laboratories for recording total ion current during use of the gas chromatographic or direct introduction inlet systems.

Most mass spectrometers utilize ultraviolet light beam galvanometers. These instruments are available with bandwidths ranging from 0 to 5000 kHz. Unfortunately, they do not provide the accuracy and linearity of servorecorders. Light beam galvanometer recorders, as used in most commercial mass spectrometers, provide accuracy and linearity of the order of 1-2%. These recorders require relatively high currents (100 μA full scale for a 1000-Hz galvanometer is typical) and, therefore, preamplifier systems are required. The dynamic range of most available recorders is quite small (about 100:1) relative to the dynamic range of mass spectrometric detectors, which may be as high as 10,000:1. This necessitates the use of multichannel recording. Usually, sensitivity ratios of 1, 10, and 100 or 1, 3, 10, 30, and 100 are used with the records offset on the chart. Most light beam galvanometer recorders operate with photographic direct printout paper and ultraviolet light beam. This provides rapid development of the record without chemical treatment. The major disadvantage of these systems is the high cost of photographic paper, which may be in the range of 30-50 cents/spectrum. This does not seem particularly high except in those situations where spectra are recorded repetitively or continually at fairly high scanning speed. At the scan rates used with modern rapid scanning mass spectrometers, it is possible to use chart paper at a rate such that the cost exceeds $1/min of machine operation. Clearly, a more economical system is required for recording the data, and magnetic tape, at least in part, fulfills this requirement (38).

C. Magnetic Tape Recorders

Magnetic tape recorders, operated in the FM mode, are capable of band widths from dc to greater than 20,000 Hz, but the bandwidth required for recording low resolution mass spectral data is of the order of 1000 Hz. A typical reel (3600 ft) of magnetic tape will record about 3 hr of continuous data at a speed of $3\frac{3}{4}$ in./sec, corresponding to a band width of 0-1250 Hz. This amount of tape costs about as much as a single 200-ft roll of photographic chart paper, which can record about 10 min of continuous data at the usual speed of 8 in./sec. The economic advantage of magnetic tape recording is clearly established. Figure 5-10 shows a mass spectrum of hexadecane recorded from a time-of-flight mass spectrometer, utilizing both the oscillographic recorder (upper trace) and the reproduction of the tape-recorded signal (lower trace). The offset between the spectra reflects the time

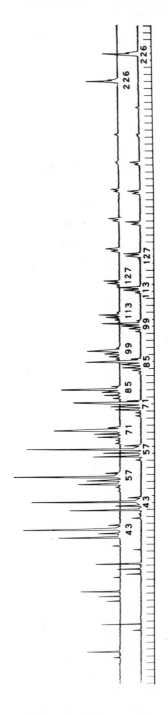

FIG. 5-10   Mass spectrum of hexadecane. Upper trace: direct
oscillographic record. Lower trace: oscillographic
record of spectrum recorded on magnetic tape (38).
(Reproduced with permission of the American
Chemical Society.)

required for the tape to move from the record to the reproduce head
of the tape recorder. The FM tape provides a completely electronic
means of recording data, and, therefore, the experiment can be
reproduced at any time for subsequent digitization or further computer
processing. Figure 5-11 shows part of the same spectrum shown in
Fig. 5-10, expanded to display a small portion of the total mass range.
This expansion illustrates the versatility of the tape recording system.
Spectra may be recorded repetitively on the tape and monitored by
playing the tape back into an oscilloscope for visual editing. Only
those portions of the tape which are of interest need be reproduced in
permanent form on an oscillographic chart record. FM tape recording
has also been shown to be an effective system for recording the electrical output of high resolution mass spectrometers for subsequent
off-line data processing (39).

D. Oscilloscope Monitoring

One other output device deserves special consideration in any complete mass spectrometric analysis system for flavor chemistry. The
oscilloscope, used as a standard output device on the time-of-flight
and quadrupole mass spectrometers, provides a very rapid indication

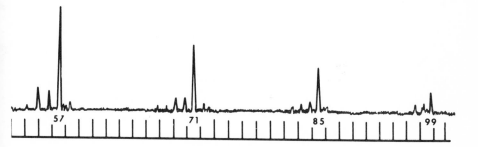

FIG. 5-11   Portion of hexadecane spectrum (m/e 57-99) recorded
            on magnetic tape at 7-1/2 i.p.s. and reproduced on an
            oscillograph. Timing signal (lower trace), 250 p.p.s.
            (38). (Reproduced with permission of the American
            Chemical Society.)

of the condition of the instrument at all times. Small changes in the
composition of the sample entering the instrument and changes in background can be monitored during repetitive scans simply by inspection
of the spectrum appearing on the oscilloscope screen. Oscilloscopes
have been used with fast scanning magnetic deflection instruments,
but one of the major problems is finding a suitable triggering system
for the oscilloscope. When magnetic field scanning is employed, the
scan is not always reproducible in time and a stable oscilloscope display is very difficult to maintain. Figure 5-12 shows an oscilloscope
triggering circuit, for use with either repetitive or single-sweep scanning (40). The primary source of the trigger signal is the magnetic
field which is measured by a Gauss meter, the output of which passes
through an adjustable threshold circuit which then triggers the oscilloscope sweep. The level of magnetic field intensity at which the oscilloscope is triggered is adjustable, so that either the entire spectrum or
small portions of it can be displayed. An oscilloscope may also be used
for monitoring of spectra reproduced from FM tape.

Figure 5-13 is a schematic diagram of the complete instrumentation
employed in one flavor research laboratory (37). The variety of inlet

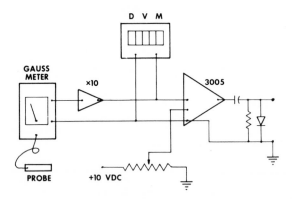

FIG. 5-12   Oscilloscope trigger circuit for magnetic deflection mass spectrometer.

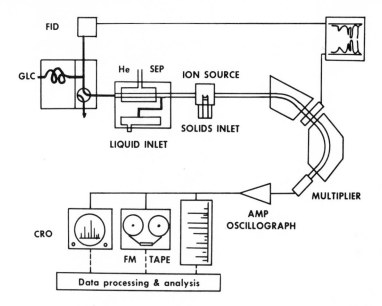

FIG. 5-13 Instrumentation for mass spectrometric analysis of flavor components (37). (Reproduced with permission of the American Chemical Society.)

systems required for the mass spectrometer (Hitachi-Perkin-Elmer, RMU-7) and the output systems available are indicated. The output amplifier is a Keithley Model 302 solid state amplifier utilizing a 10^{9} - Ω input resistor. This provides an overall bandwidth of 1000 Hz for all of the output systems. The total ion current monitor is located between the electrostatic and the magnetic sectors of the mass spectrometer. At this point, because of mass discrimination in the ion source and some deflection in the electrostatic sector, the helium ions from the carrier gas can be eliminated from the total ion current signal.

Note that all of the inlet systems described above are included, and each of these has its place and has been used in flavor investigations.

VII. SAMPLE SIZE REQUIREMENTS

The problem of defining sample size requirements for identification of organic compounds isolated in flavor investigations can become very complicated. The amount of sample required depends on the nature of the material and its decomposition mechanism, that is, whether a small number or a large number of ion fragments is produced. The more fragment ions produced, the more sample required for identification. Different methods of introduction provide different detection and identification limits in qualitative mass spectrometry. McFadden (10) has indicated detection limits of 1, 10^{-3} to 10^{-4}, 10^{-3} to 10^{-4}, 10^{-3}, and 10^{-4} µg for samples introduced into the mass spectrometer by a typical batch inlet, a small volume batch inlet, direct introduction probe, packed column GC plus separator, and capillary GC systems, respectively. Identification limits are stated to be approximately 100 times higher; that is, 100 times more sample is required for identification by mass spectrometry than for detection. In general, these values are realistic, though some variation according to the type of sample may be expected. High quality spectra can usually be obtained with sample quantities of the order of 1 µg. Sample quantities in the range of 1-10 µg will provide enough material either for repeated analysis by using the gas chromatographic inlet system or for prolonged examination of the sample at different instrument sensitivities so that metastable peaks can be recorded. High resolution spectra can be recorded if necessary, and precise isotope peak intensity measurements can be made. In general, sensitivity limits are set not by the amount of sample but by the amount of contamination present, either as instrument background or as contaminant introduced with the sample.

VIII. ARTIFACTS

There is some problem of artifact production during sample collection, transfer, and final analysis by mass spectrometry. The major problem is that of thermal decomposition in the inlet system. This can be largely avoided by use of the direct inlet, which permits the use of temperatures 100-200° C lower than those that must be used when other inlet systems are employed. The other major source of artifacts in mass spectral examination of components of food flavors are those that arise from bleeding gas chromatographic columns or impurities in thin layer plates and solvents. These can normally be avoided by careful sample preparation and attention to details. Occasionally, sollision-induced reactions at high source and analyzer pressures can result in distorted mass spectral patterns; therefore, efficient pumping systems are required, preferably differential pumping for the ion source and analyzer.

IX. SUMMARY

Mass spectrometry provides the highest sensitivity of any of the available spectrometric methods for identification of compounds which may be isolated during the course of flavor investigations. However, mass spectral data are of most value when interpreted in the light of other information. Gas chromatographic data are usually available and may provide some clues for identification. Nuclear magnetic resonance and infrared spectra can be particularly useful in aiding interpretation of mass spectra. Even if only very small quantities of sample are available and high quality NMR and IR spectra cannot be obtained, it is still sometimes possible to get some information by these methods. For example, the presence of a carbonyl group may be indicated by the infrared spectrum of a sample as small as a few micrograms. This kind of information will aid greatly in interpretation of the mass spectrum.

A large variety of mass spectrometry instrumentation systems is available. The choice of instrument depends primarily on the kinds of samples to be handled and the budget available. The most important and often most difficult problem is getting the sample into the instrument. Therefore, it is essential that all inlet systems be available.

The best general approach to mass spectrometric analysis of an unknown sample is to get as much data as necessary. A normal low
resolution fast scan survey spectrum should be recorded. If unambiguous identification can be made, no further effort is necessary. If
the sample is impure and there is some doubt about which peak is the
molecular ion, spectra can be recorded at low electron energy to
determine which peak is the molecular ion of each component. If no
molecular ion appears in the 70-eV spectrum, it is unlikely that one
will be observed at low voltages.

High resolution spectra may be recorded to help settle ambiguities.
These are not generally required for the majority (probably more than
90%) of the compounds encountered in flavor investigations. Therefore, it is generally not necessary for a laboratory which is involved
primarily in flavor studies to have continuous access to a high resolution mass spectrometer. High maintenance expenses and operational
difficulties associated with high resolution mass spectrometers make
them less desirable as routine instruments in the flavor laboratory
than the conventional medium resolution, general-purpose instruments.
Occasional access to a high resolution instrument is usually adequate.
Generally, it is not desirable to employ a high resolution machine for
work in which the majority of the spectra are recorded at low resolution. High resolution instruments have minimum resolving powers in
the range of 1000-2500, determined by the presence of fixed inter-
mediate slits, and are therefore less sensitive than low resolution,
general-purpose instruments.

The mass spectrometer has become, in recent years, an indispensable tool to the flavor chemist. Instrument selection has become a
problem because of the wide variety of fast scanning, high sensitivity,
general-purpose instruments presently commercially available. All
mass spectrometrists have their biases and prejudices for particular
types and brands of instruments, and those flavor chemists seeking
advice as to which instrument to buy usually find little concurrence

among professional mass spectrometrists. Actually, most of the
available commercial instruments are rugged, dependable, and satisfactory for the majority of problems that must be solved in the flavor
laboratory. The most important variable is still the instrument operator and user. Most chemists can adapt their experimental methods to
the capabilities of the available instrument and utilize it for solution
of their problems.

REFERENCES

1. G. R. Hercus and J. D. Morrison, Australian J. Sci. Res.,
    B-4, 290 (1951); A. Turk, R. M. Smock, and T. I. Taylor,
    Food Technol., 5, 58 (1951).

2. W. H. Stahl, Quartermaster Food and Container Inst., Surveys
    of Progress on Military Subsistence Problems, Ser. 1, No. 9,
    1957, p. 58.

3. K. Biemann, Mass Spectrometry: Organic Chemical Applica-
    tions, McGraw-Hill, New York, 1962.

4. H. Budzikiewicz, C. Djerassi, and D. H. Williams, Mass
    Spectrometry of Organic Compounds, Holden-Day, San
    Francisco, 1967.

5. F. W. McLafferty, Interpretation of Mass Spectra, W. A.
    Benjamin, New York, 1966.

6. E. Stenhagen, S. Abrahamsson, and F. W. McLafferty, eds.,
    Atlas of Mass Spectral Data, Wiley (Interscience), New York,
    1969.

7. H. J. Beynon, Mass Spectrometry and Its Applications to
    Organic Chemistry, Elsevier, Amsterdam, 1960.

8. J. Roboz, Introduction to Mass Spectrometry. Instrumentation
    and Techniques, Wiley (Interscience), New York, 1968.

9. R. Ryhage, Arkiv. Kemi, 20, 185 (1962).

10. W. H. McFadden in Advances in Chromatography (J. C. Giddings
    and R. A. Keller, eds), Vol. 4, Dekker, New York, 1967,
    p. 265.

11.   C. Merritt, Jr. , M. L. Bazinet, J. H. Sullivan, and D. H.
      Robertson, J. Agr. Food Chem. , 11, 152 (1963); M. L. Bazinet
      and C. Merritt, Jr. , Anal. Chem. , 34, 1143 (1965).

12.   H. J. Svec and D. D. Clyde, J. Chem. Eng. Data, 10, 151 (1965).

13.   K. Biemann, C. Cone, B. R. Webster, and G. Arsenault,
      J. Am. Chem. Soc. , 88, 5598 (1966).

14.   M. Senn, R. Venkataraghavan, and F. W. McLafferty, J. Am.
      Chem. Soc. 88, 5593 (1966).

15.   J. W. Amy, E. M. Chait, W. E. Baitinger, and F. W.
      McLafferty, Anal. Chem. , 37, 1265 (1965).

16.   J. N. Damico, N. P. Wong, and J. A. Sphon, Anal. Chem. ,
      39, 1049 (1967).

17.   M. H. Studier, 11th Ann. Conf. Mass Spectrometry and Allied
      Topics, San Francisco, 1963, p. 142.

18.   H. Tsuyama, H. Hirose, and Y. Nakajima, 15th Ann. Conf.
      Mass Spectrometry and Allied Topics, Denver, 1967, p. 532.

19.   J. Okamoto and E. Mitani, 15th Ann. Conf. Mass Spectroscopy
      and Allied Topics, Denver, 1967, p. 526.

20.   R. W. Kiser, Introduction to Mass Spectrometry and Its Appli-
      cation, Prentice-Hall, Englewood Cliffs, N. J. , 1965, p. 59.

21.   W. K. Rohwedder, 15th Ann. Conf. Mass Spectrometry and
      Allied Topics, Denver, 1967, p. 429.

22.   J. N. Damico, R. P. Barron, and J. M. Roth, 15th Ann. Conf.
      Mass Spectroscopy and Allied Topics, Denver, 1967, p. 433.

23.   H. D. Beckey, H. Knoppel, G. Metzinger, and P. Schulze,
      Advances in Mass Spectrometry (W. L. Mead, ed.), Vol. 3,
      The Institute of Petroleum Press, London, 1966, p. 35.

24.   J. Block, Advances in Mass Spectrometry (E. Kendrick, ed.),
      Vol. 4, The Institute of Petroleum Press, London, 1968,
      p. 791.

25.   F. H. Field, Advances in Mass Spectrometry (E. Kendrick, ed.),
      Vol. 4, The Institute of Petroleum Press, London, 1968,
      p. 645.

26. F. J. Comes, Advances in Mass Spectrometry (E. Kendrick, ed.), Vol. 4, The Institute of Petroleum Press, London, 1968, p. 739.

27. I. Howe and D. H. Williams, Chem. Commun. 1968, 220.

28. H. W. Major, Perkin-Elmer Corp., personal communication, 1968.

29. R. Ryhage and S. Wickstrom, Science Tools, L. K. B. Instruments, 14, 1 (1967).

30. G. L. Kearns, Picker Corp., personal communication, 1969.

31. W. G. Nihaus, Jr., and R. Ryhage, Tetrahedron Letters, 49, 5020 (1967).

32. R. S. Gohlke, Anal. Chem., 31, 535 (1959).

33. W. C. Wiley and I. H. McLaren, Rev. Sci. Instr., 26, 1150 (1955).

34. R. D. Craig, B. N. Green, and J. D. Waldron, Chimia, 17, 33 (1963).

35. K. Biemann, P. Bommer, D. M. Desiderio, and W. J. McMurray, Advances in Mass Spectrometry (W. L. Mead, ed.), Vol. 3, The Institute of Petroleum Press, London, 1966, p. 639.

36. A. O. Nier and E. B. Delany, 15th Ann. Conf. Mass Spectrometry and Allied Topics, Denver, 1967, p. 484.

37. P. Issenberg, A. Kobayashi, and T. J. Mysliwy, J. Agr. Food Chem., 1969, 17, 1377 (1969).

38. P. Issenberg, M. L. Bazinet, and C. Merritt, Jr., Anal. Chem., 37, 1074 (1965).

39. C. Merritt, Jr., P. Issenberg, M. L. Bazinet, B. N. Green, T. O. Merron, and J. G. Murray, Anal. Chem., 37, 1037 (1965).

40. P. Issenberg, Massachusetts Institute of Technology, unpublished, 1968.

Chapter 6

COMBINED GAS CHROMATOGRAPHY-MASS SPECTROMETRY
(GC-MS): SPECIAL TECHNIQUES, DATA PROCESSING

I. COMBINED GC-MS

A. Alternative Inlet Systems

Maximum information can be obtained from mass spectra when the slowest scan rate consistent with the rate of sample consumption is employed. At slow scanning rates, the bandwidth of the recording system can be reduced by filtering and sensitivity can thereby be improved by integrating the statistical fluctuations in ion current. Mass spectral fragmentation patterns are more reproducible at low scan rates. Current sensitivity can be improved by using higher amplifier input resistances. Resolving power and apparent sensitivity of a mass spectrometer are degraded by increasing the scan speed so that peak width becomes comparable to the time constant of the recording system (1). These limitations are not approached by most commercial instruments employing signal bandwidths greater than 1000 Hz at scan times in the range of 1-10 sec/decade. The useful dynamic range of the mass spectral signals is decreased by rapid scanning (2), and very small peaks, particularly those with relative intensities of less than 1%, may be lost or measured with poor precision. Dynamic range may be significantly limited when the amounts of sample examined are very small.

Metastable peaks provide a great deal of useful structural information (3) and are especially valuable in interpretation of spectra of unusual compound types and new structures. These peaks are usually relatively noisy and of low intensity, generally less than 1% of the most intense peaks, in spectra recorded under normal or fast scan conditions. Methods have been developed for examination of metastable transitions and for unambiguous determination of mass of the parent and daughter ions by using the energy-focusing property of double-focusing mass spectrometers (4). Two approaches have been used to study the metastable transitions occurring in the region between the ion source and the electrostatic sector of the double-focusing mass spectrometer. The ratio of accelerating voltage to electrostatic analyzer voltage can be varied either by increasing accelerating voltage at constant electrostatic sector voltage (5), or by decreasing the electrostatic analyzer voltage while maintaining constant accelerating potential (6). Both methods defocus the "normal" spectrum, providing peaks which represent only the metastables with kinetic energy lower than that of normal ions. The concept of "metastable maps" (7) can be used to aid in the elucidation of complex molecular structures. Application of this approach requires slow scanning and multiple scans in order to record all of the required data.

Collection of the maximum amount of information from mass spectral analysis in a difficult structural problem would require introduction of the sample through a conventional reservoir-type batch inlet system, slow scanning of a spectrum, and repeated scanning of the spectrum at progressively lower electron energies to obtain appearance potential data. This would be followed by appropriate metastable enhancement techniques for those transitions of interest. High resolution data, yielding accurate mass measurements, may be required. This kind of examination requires long time expenditures for each sample and often is not justified by the complexity of the structure, particularly in examination of the multicomponent mixtures encountered in flavor studies. Though most of the mixtures contain many components (hundreds are not uncommon), most of these are relatively simple compounds which can be identified rapidly by examination of a conventional low resolution mass spectrum.

B. Advantages of Combined GC-MS

The principal advantage of the gas chromatographic inlet system, with continuous mass spectrometric monitoring of the column effluent, is convenience. The gas chromatographic inlet is the simplest inlet system to operate. Anyone skilled in operation of a conventional gas chromatograph can inject a sample and push the mass spectrometer scan button at the appropriate time. Manipulation of the complex

valving and pumping systems required for introduction via the other
inlet systems is not necessary. Sample loss and contamination,
which usually accompany trapping of gas chromatographic peaks, are
eliminated entirely. The sample is delivered to the mass spectrometer ion source at the maximum purity possible. Many spectra can
be recorded from individual peaks as they are eluted from the gas
chromatographic column. Differences between spectra recorded at
the beginning, middle, and end of the gas chromatographic peak often
provide information which will aid in detection and identification of
incompletely resolved components. The presence of column bleed in
trapped gas chromatographic fractions is always a problem when the
fractions are examined by mass spectrometry. The longer the trapping time, the greater will be the contamination of the sample by
liquid phase. This problem has been alleviated through use of lower
volatility silicones with a wide range of polarities (the OV-series of
liquid phases), but column bleed continues to be a problem in mass
spectrometric examination of samples collected from gas chromatographic columns. When a gas chromatographic inlet is used, column bleed can be monitored continually between peaks and more
readily subtracted from spectra of resolved components.

The mass spectrometer is sensitive to all compounds of sufficient
volatility. When conventional batch-type inlet systems are employed,
solvents cannot be used for sample transfer. When the gas chromatographic inlet is employed, solvents can be used for convenient
transfer of trapped fractions. Solvents are easily removed from
samples on even a relatively inefficient gas chromatographic column.

C. Scan Rate and Sensitivity Requirements

In order that mass spectra recorded with the GC-MS system be
reproducible and typical of the mass spectral pattern of the compound
being examined, as well as being comparable to spectra recorded
using alternative inlet systems, it is necessary that a relatively constant
partial pressure of sample in the ion source be maintained during the

time required to record the spectrum. Gas chromatographic peak widths and the rate of change of sample concentration in the carrier gas vary over wide ranges, depending on the type of column employed and the retention time of the component. When high efficiency columns are used, peak widths in the range of 10-30 sec are common and mass spectra must be scanned within 1-5 sec if reproducible, unbiased spectra are to be recorded. Minimum rate of change of sample concentration in the carrier gas is observed at the top of a chromatographic peak. It is in this region that the most reproducible mass spectra can be recorded.

Of all spectrometric methods available, mass spectrometry most closely approaches the sensitivity of conventional gas chromatography ionization detectors. Sample consumption rates of 10^{-9} g/sec can give excellent mass spectra. "Useful" information can be obtained at sample consumption rates as low as 10^{-11} g/sec under optimum conditions.

Determination of the actual useful sensitivity of a combined GC-MS system is a complex problem, particularly when attempting to compare results of different investigators or to evaluate specifications of different instrument manufacturers. It is impossible to specify sensitivity in terms of total amount of sample injected on a gas chromatographic column. Peak widths vary greatly with column efficiency and retention time. Therefore, sample consumption rate (grams per second) must be specified. * Useful sensitivity will vary with the type of sample examined. Though total ion current is comparable for most organic compounds of similar molecular weight, the intensity of important peaks in the spectrum will vary greatly depending upon the fragmentation pattern. A given current measured at the total ion current monitor may give rise to a spectrum containing only one or two major peaks. In this case, effective sensitivity will be extremely high. If a large number of fragment peaks of relatively low intensity are produced, useful sensitivity is diminished.

* In addition, the resolving power at which the sensitivity is measured must be stated.

Sensitivity is usually specified in terms of signal-to-noise ratio (S/N).
Conventionally, a value of S/N = 2 is taken as a limit of detection. At
such low signal-to-noise ratios, it is unlikely that useful spectra can
be recorded. A more reasonable value would be a signal-to-noise
ratio of approximately 10 (the ion fragment for which this detection
limit is specified must also be indicated). One must state the nature
of the noise included. Electron multipliers are capable of detecting
single ions arriving at the detector. It is necessary to specify whether
the noise level corresponds to the signal observed for a single ion
arriving at the multiplier or to the overall electrical noise level in the
amplifying and recording systems. Both are important under different
operating conditions.

Instrument manufacturers' specifications usually refer to optimum
operating conditions for new instruments with clean ion sources,
which have been properly aligned and adjusted to deliver maximum
sensitivity. This performance should not be expected in routine
laboratory operations when the source is dirty, background contamination is high, or column bleed interference is significant. Conditions
peculiar to individual laboratories and installations may also reduce
the practical sensitivity limits of an instrument. Problems which
have been encountered include excessive building vibration, stray
magnetic fields in the laboratory, and high frequency noise on the ac
power supply.

D. GC-MS Interfacing—General Considerations

The principal difficulty in connecting a gas chromatographic inlet
system to the mass spectrometer is the necessity for maintaining high
vacuum in the mass spectrometer even though a continuous flow of
carrier gas is admitted from the gas chromatographic column. Carrier gas flow rates are usually in the range of 10-60 cc/min for small
diameter packed columns and 0.5-10 cc/min for open tubular columns.
Mass spectrometer pumping systems usually have capacities in the
range of 0.1-0.3 cc (STP)/min to maintain analyzer pressures in the

range of 10^{-6} to 10^{-5} Torr, even when differential pumping is used.
At high analyzer pressures, pressure broadening results in the loss
of effective resolving power, and changes in mass spectral fragmentation patterns may be observed (8). The latter effect may be due to
ion-molecule reactions in the analyzer. This effect can be reduced by
differential pumping.

A practical point, often overlooked, is that most vacuum gauges in
use on mass spectrometers are calibrated for air or nitrogen. Typical ionization vacuum gauges are about five times less sensitive to
helium than to nitrogen. There may be a large discrepancy between
indicated and true pressure.

It is necessary to consider all of the plumbing problems and the
special considerations required for ordinary gas chromatography in
the design of GC-MS connection systems. Length of connecting tubing
must be kept to a minimum. The diameter of the tubing should be kept
as small as possible, using capillary tubing whenever feasible. Careful temperature control of all connecting lines, valves, and accessories
is essential. One must always be aware of the possibility of thermal
degradation of sensitive compounds between the end of the gas chromatographic column and the mass spectrometer ion source. Unswept
volumes on the low pressure side of the connection system must be
kept to an absolute minimum.

E.   Column Effluent Splitters

The simplest apparatus for connecting the outlet of a gas chromatographic column to a mass spectrometer ion source consists of a
"T" connection placed at the end of the column. One of the exit lines
is connected by means of a flow-control needle valve to the ion source.
An auxiliary pumping system is connected to the third port to remove the
excess sample. This system is appropriate with 0.01-in.-i.d. open
tubular columns, since at the low flow rates which are optimum for
these columns between 25 and 100% of the column effluent can be

introduced into the mass spectrometer without overloading the source
pumping system (9, 10). If the column pressure drop is greater than
1 atm, the head pressure is simply reduced by 15 psi to compensate
for the fact that part of the column is operated under vacuum. Retention times and column efficiency may be maintained equal to those
observed when the column is operated with the outlet at atmospheric
pressure (11). When packed gas chromatographic columns are employed, the proportion of sample entering the spectrometer is significantly smaller. Gohlke (12) reported that at a carrier gas flow rate
of 60 cc/min approximately 0.3% of the column effluent passed into
the mass spectrometer. Recoveries this small may be adequate when
sample quantities are in the milligram range, as is the case in
examination of many synthetic reference compounds. A fundamental
advantage to using only a small portion of the sample in the mass
spectrometer is that the remaining sample (greater than 99%) may be
recovered and examined by other methods.

F. Carrier Gas Separators

To maintain maximum sensitivity, it is necessary to provide a
system which will preferentially remove a large proportion of the
carrier gas while permitting sample components to enter the ion
source. Helium is commonly used as a carrier gas. Devices of
varying effectiveness and complexity have been described which utilize
the great differences in molecular properties of helium and organic
compounds. Most utilize the differences in rates of diffusion under
conditions of molecular flow or differences in solubility in polymeric
materials. They provide enrichment ratios, i.e., increases in concentration of organic compounds, between 10 and 100. In addition to
enrichment ratios, it is necessary to consider the efficiency of such
a device. Efficiency of a carrier gas separator is defined as the
fraction (or percentage)

$$\text{separator efficiency} = \frac{\text{g/sec entering ion source}}{\text{g/sec entering separator}} \qquad (1)$$

Carrier gas separators in common use are illustrated schematically in Fig. 6-1. Figure 6-1a is a diagram of the Watson-Biemann separator (13). The carrier gas (helium) plus sample passes through a cylindrical fritted glass tube which has an average pore diameter of 1μ or less. The helium, because of its low molecular weight, preferentially diffuses through the fine pores; some of the solute also passes through the fritted glass, but the result is a relative enrichment of the sample concentration in the carrier gas. For diethyl ether, an enrichment of about 50-fold was observed with 10-50% efficiency. By optimizing flow restriction at the ends of the separator, particularly the restriction between the separator and the ion source, R_2, efficiencies of over 40% can be consistently attained for n-hexane eluted from a packed column at a helium flow rate of 15 cc/min (14). Grayson and Wolf (14) employed a variable restrictor at R_2, in the form of a needle valve, to optimize efficiency for different carrier gas flow rates. A two-stage separator of the Watson-Biemann type has been described (15) which affords greater enrichment at high carrier gas flow rates. To reduce dead volume which could result in peak broadening, a separator was constructed with a cylindrical glass rod placed within the hollow fritted tube. This modification resulted in retention of column efficiency when open tubular columns were employed with the diffusion-type separator (15). As solute molecular weight increases, greater enrichment and efficiency can be expected for all diffusion-type separators. Some problems resulting from adsorption of polar solutes on the large glass surface of the Watson-Biemann-type separator have been reported, but these can be greatly diminished by conventional silanizing techniques, which are familiar to most chromatographers (15). Treatment with bis-trimethyl silylacetamide (BSA) resulted in increases of sample sensitivities of two and three orders of magnitude for terpene esters and alcohols, respectively (16). The Watson-Biemann-type separator is probably the most widely used in combined GC-MS. It is inexpensive (less than $50), and its all-glass construction makes it relatively inert

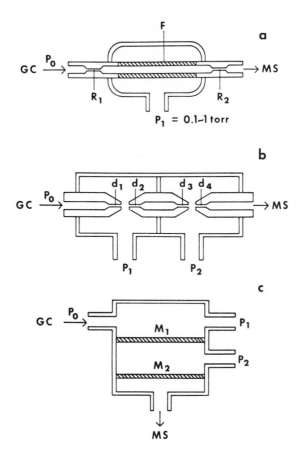

FIG. 6-1   Carrier gas separators for GC-MS. a, Watson-
            Biemann separator (13), b, Ryhage separator (19),
            c, Llewellyn separator (20).

except for the adsorption problems already described. A similar device has been constructed using a porous silver diffusion medium (17),
which offers some advantages by reducing dead volume and adsorption
surface area.

Figure 6-1b shows, in schematic form, the carrier gas separator
described by Ryhage (18). Effluent from the GC column enters the

separator at atmospheric pressure through a very fine jet at greater than sonic velocity. Heavier molecules are enriched in the stream entering d_2. Helium diffuses away at a greater rate than heavier solute molecules. The first stage reduces the pressure to about 0.1 Torr at a helium flow rate of 30 cc/min. Most of the remaining carrier gas is removed in the second stage of the separator by passage of the gas through the orifices d_3 and d_4. All orifices are between 0.1 and 0.3 mm in diameter. Source pressure can be maintained at about 1.2×10^{-6} Torr at 30 cc/min helium flow rate, measured at the column outlet. Efficiency of about 55% was measured for tetramethylpyrazine (mol. wt. 136) (19). High velocity, in the molecular beam produced, should ensure minimal loss of chromatographic resolution; however, no evaluation of the performance of the separator with very high efficiency open tubular columns has been reported.

The partition separator described by Llewellyn and Littlejohn (20) is shown schematically in Fig. 6-1c. Separation is effected by a membrane in which the organic sample is more soluble than is the carrier gas. Silicone rubber membranes exhibit selectivity for organic compounds, and a two-stage separator can provide 90% efficiency with almost complete rejection of carrier gas. Selectivity of the membrane varies with membrane composition, solute polarity, and operating temperature. The time constant is probably greater than for the other two separators described, and could be expected to result in some loss of chromatographic resolution, especially with open tubular columns. The separator behaves, in a sense, as a single-plate partition column. Black et al. (21) reported efficiencies between 30 and 50% with a single-stage version of the Llewellyn separator. The portion of column effluent not admitted to the mass spectrometer ion source passed through a flame ionization detector. Indication of the chemical selectivity of this membrane separator is shown in Fig. 6-2 (21). The discrimination observed is probably no greater than that exhibited by other types of separators and is predictable on the basis of chromatographic behavior on columns prepared from material similar to

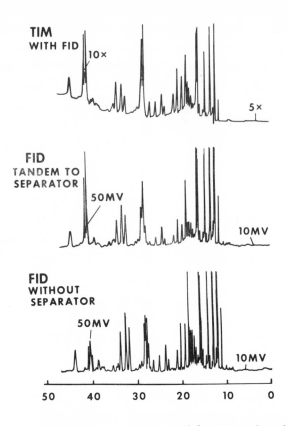

FIG. 6-2   Chromatogram of orange oil fraction using single-
stage phenylmethylsiloxy copolymer membrane
separator and 0.03-in.-i.d. x 500-ft OV-101 column (21). Reproduced with permission of the
Journal of Chromatographic Science.

that used in membrane preparation. The influence of the separator on

the efficiency of open tubular columns, which exhibited greater than

100,000 theoretical plates, was evaluated. In all cases band spreading

was observed, but the device was still considered extremely useful in

flavor studies. It is unfortunate that studies have not been made of

the influence of other types of separators on spreading of bands from

such high resolution columns. The ability to operate the membrane separator with the waste sample being vented to atmospheric pressure provides a major advantage in flavor investigations. Since the excess gas is not evacuated through a vacuum system, it is possible to smell the column effluent and to use odor as a guide for determining the appropriate time to record spectra.

It is likely that the single-stage version of the silicone membrane separator will find much wider application in the future. The effluent from this separator may be "sniffed", and collection devices may be used to trap fractions.

G. Chromatographic Detectors

Most mass spectrometers are equipped with total ion current (TIC) monitors which can serve as chromatographic detectors during GC-MS operation; however, there are advantages to using auxiliary detectors in parallel with the mass spectrometer. Usually the end of the column is separated from the mass spectrometer interface by a length of capillary tubing which provides a transport time lag between appearance of sample components at the column outlet and their appearance in the ion source. Leemans and McCloskey (8) used this time lag to advantage for permitting the mass spectrometer operator to determine, from the preview chromatogram measured by the chromatographic detector, the optimum time at which to scan the mass spectrum. They showed that the time lag could be varied conveniently, when packed columns were employed in the GC-MS system, to provide the operator with a chromatographic record giving a lead time of 4-6 sec before the sample appeared in the mass spectrometer.

Use of the auxiliary chromatographic detector with a combined GC-MS system affords a means for rapid check on operation of the system. Thermal decomposition and adsorption in the connecting lines, separator, or pressure reduction system can be detected by noting discrepancies between chromatographic records recorded by the auxiliary detector and by the TIC. Microthermal conductivity

detectors could be used for this application. They offer the advantage
that no splitter is required, and they can be used in series with the
chromatographic column and mass spectrometer.

The most common arrangement of an auxiliary detector in the
GC-MS system employs the flame ionization detector with a splitter at
the column outlet (see Fig. 5-13). Under the most favorable conditions, sensitivity of the mass spectrometer is comparable to or greater
than that of a flame ionization detector. Split ratios are usually selected to transport a larger proportion of the column effluent to the mass
spectrometer than is consumed in the flame ionization detector. Sample loss in the particular interfacing system employed must also be
considered in establishing the optimmu FID-MS split ratio. In most
GC-MS interfaces that have been described, the connecting line between
the chromatographic column and the interface is maintained under
vacuum. The requirement for maintaining atmospheric pressure at
the flame ionization detector will also influence the available range of
split ratios.

The TIC monitor is the most appropriate detection system for measuring sample concentration entering the mass spectrometer from a
gas chromatographic column. There is no significant time lag between
detection of ions by the TIC monitor and their arrival at the detector
of the mass spectrometer. It is usually possible to establish consistent relationships between the signal from the total ion current monitor
and the minimum detectable amounts for which spectra can be
recorded.

The time-of-flight mass spectrometer employs a unique TIC
monitoring system. In this instrument, total ion current is measured
at the multiplier by integrating all ions from a fixed proportion
(between 10 and 50%) of the total spectra scanned. Every tenth (10%)
or alternate (50%) cycle is gated from the multiplier to the monitor
amplifier. Since 10,000 spectra/sec are commonly scanned on this
instrument, the TIC signal appears to be a dc signal on a conventional

strip chart servopotentiometer. No interruption or transient disturbance is produced in other output systems. Current from helium ions is
eliminated from the TIC by applying a pulse which effectively turns
the multiplier off during the early part (through m/e 12) of each cycle.

In the quadrupole mass spectrometer, the TIC monitor integrates
multiplier output, an approach similar to that employed in the time-of-
flight instrument. The result is similar when repetitive scanning at
high rates (100 scans/sec) is employed. However, at the reduced
scanning rates required for recording on an oscillographic recorder
(1-10 sec), the auxiliary spectrum recording system causes a transient
disturbance in the TIC integral recording.

In a magnetic deflection mass spectrometer the TIC monitor consists of an electrode, located either in the ion source or at some
point between the source and the magnetic analyzer, which intercepts
a fixed fraction (between 10 and 50%) of the unresolved beam. Sensitive electrometer amplifiers, employing high impedance input circuits,
are required to measure the total ion current. When the magnetic
field is scanned at high speeds, a transient deflection of the TIC signal is
usually observed. This disturbance is probably due to small currents
induced in the electronic circuitry by rapid changes in the magnetic
field near them. Another disadvantage of these monitors is that all
ions are recorded, including those contributed by the carrier gas.
Unless very efficient separating devices are employed, excessive
background signal is contributed by the carrier gas, resulting in the
necessity for use of high offset voltages in the monitor amplifier.
Though this constant background current due to carrier gas may be
effectively "balanced" in the amplifier, there is a resultant increase
in noise level in the TIC record which may reduce the effective
sensitivity of the combined GC-MS system.

One approach to elimination of the contribution of carrier gas
ionization to the TIC record takes advantage of the large difference in
ionization potentials of helium and most organic compounds. The

ionization potential of helium is 24.5 eV. If the ion source is operated with an electron energy of 20 eV, helium is not ionized and the TIC record represents only organic compounds (18). Mass spectra recorded at 20-eV electron energy are very similar, but not identical, to those recorded at 70 eV. Brunee et al. (22) described a double ion source, in which the entering gas sample is split in two portions. One portion passes into a conventional ion source operated at 70 eV, and the ion beam produced is analyzed and recorded in the conventional manner using a magnetic sector instrument. The other half of the sample enters an ion source operated at 20 eV. The ion current is collected on an electrode, amplified, and displayed as the total ion current, with no contribution from helium. The 20-eV portion of the ion source is located far enough away from the magnetic field so that no transient disturbance of the total ion current record is produced when a spectrum is scanned. Spectrum bias, resulting from sample pressure changes while scanning, is eliminated by a circuit which divides the multiplier output by the monitor output prior to recording.

In double-focusing instruments, the TIC monitor is usually located between the electrostatic and magnetic analyzers. At this point there is some dispersion of the ion beam, caused in part by the collimating magnet in the ion source and in part by deflection in the electrostatic sector. One can take advantage of this slight dispersion. With careful adjustment of source operating potentials, electrostatic sector potentials, and TIC monitor position, a signal can be recorded that includes only those ions with masses greater than that of helium.

H. Application to Flavor Problems

Figure 6-3 (23) is a chromatogram recorded simultaneously by a flame ionization detector placed at the column outlet and the TIC monitor of a double-focusing mass spectrometer. The sample was prepared from the neutral fraction of a hydrolyzed banana aroma concentrate. It is a fraction from a 1/4-in. Carbowax 4000 column. The

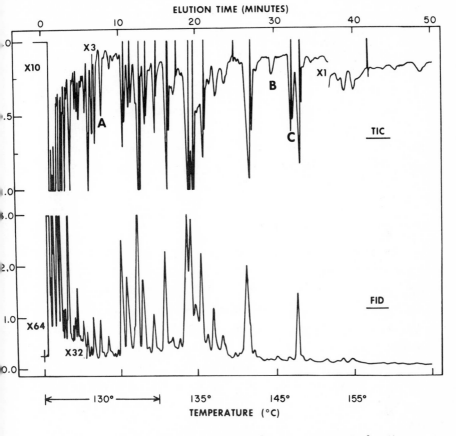

FIG. 6-3   Chromatogram of banana flavor concentrate fraction:
0.02-in.-i.d. x 50-ft Carbowax 1540 SCOT column.
TIC: mass spectrometer total ion current record.
FID: flame ionization detector record (23).
Reproduced with permission of the Institute of Food
Technologists.

sample was then eluted from a packed SE-30 column to remove
silicone impurities introduced during preliminary processing. The
inlet chromatograph of the mass spectrometer contained a 50 ft x
0.02 in. Carbowax 1540 support-coated open tubular column. Flow
rate was 10 cc/min, providing a transport lag of about 3 sec between
the signal indicated by the FID and that indicated by the TIC monitor.

The ratio of sample flow through the FID to that entering the GC-MS interface was about 1:4. The interface consisted of a stainless steel capillary restriction (6 ft long by 0.01 in. i.d.) attached to a Watson-Biemann-type separator. The entire system was silanized to minimize adsorption of polar compounds. The chromatogram in Fig. 6-3 indicates that no significant loss of chromatographic resolution occurred during passage through the connecting lines, switching valve, or helium separator. The TIC record shows the transient disturbances when the magnetic field is swept to record a spectrum. The FID, of course, is not affected. Valid chromatographic data may be acquired, even during repetitive magnetic scanning, when an auxiliary detector is employed.

Though the efficiency of the separator was only about 20%, sensitivity was adequate for most purposed. The peak at 7.5 min (A) represents approximately 0.1 µg of sample injected. Two significant discrepancies appear in the TIC and the FID records. The peaks eluted at 29 (B) and 32 (C) min are silicone impurities to which the FID is relatively insensitive, while the TIC monitor exhibits uniform sensitivity to essentially all compounds. The extent to which fractions, collected from silicone columns, were contaminated with low molecular weight silicone compounds was not realized until the mass spectrometer was employed. Fortunately, the unique distribution of the naturally occurring stable isotopes of silicone makes them easy to detect and identify in mass spectra.

Because of the complexity of mixtures of volatile materials isolated from natural food samples, high efficiency chromatographic columns are essential for their separation. Use of these columns imposes stringent sensitivity requirements on the identification methods which are later employed, since high separation efficiency is attained only for small samples. The maximum amount which can be applied to high efficiency columns is limited by the amounts of major components. Columns will be damaged by massive overloading with sample or solvents.

Figure 6-4 shows separation of an aroma concentrate, prepared from bananas, on a 500 ft x 0.02 in. SF-96 (50) open tubular column, programmed from 75 to 175° C (24). Many peaks represent several components, but the chromatogram is probably a realistic representation of the complexity of ripe banana odor concentrate. We should examine this chromatogram from the point of view of the performance required of a GC-MS system. The major components are present in microgram quantities. For example, the peak eluted at 20 min (A) represents approximately 18 µg. This approximation is based on calibration of the GC detector with n-amyl butyrate, as are all the quantities stated. Mass flow rate of sample at the peak maximum is 1.6 µg/sec. The peak at 43 min (B) represents about 1.5 µg with a maximum mass flow rate of 0.10 µg/sec. The peak at 110 min (D) contains 0.23 µg with a mass flow rate of about 12 ng/sec. The peak at 101 min (C), one of the smallest detectable in this chromatogram, contains about 20 ng; mass flow rate at the peak maximum is of the order of 0.9 ng/sec. This single chromatogram exhibits a range of component quantity greater than 1000:1. The corresponding mass flow rate range is 1800:1. Most instruments are capable of recording mass spectra of sufficient quality to provide some structural information even for the smallest peaks in this chromatogram. The problem is not in ultimate sensitivity of the mass spectrometer but rather in the dynamic range of sample flow rates that is encountered in aroma concentrates. It is not possible to use a single set of instrument operating parameters to record highest quality spectra for all of these components.

Another significant feature illustrated by Fig. 6-4 is the column bleed observed at the higher temperatures. At 175° C, the base line signal is equivalent to that produced by 4.2 ng/sec of amyl butyrate. It is approximately five times the signal measured at the top of the peak eluted at 101 min (peak C). Since this bleed consists primarily of silicone compounds, to which the FID is relatively insensitive, the bleed rate is actually higher. Silicone compounds are major

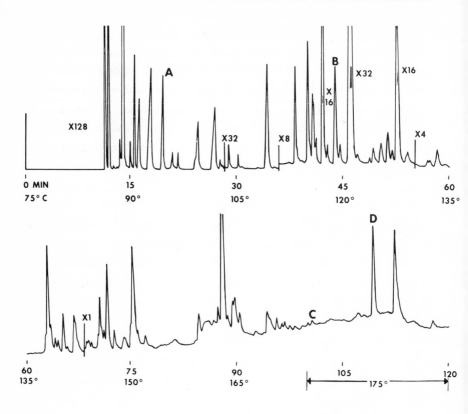

Peak	Amount, µg	Maximum mass flow rate µg/sec
A	18	1.6
B	1.5	0.1
C	0.02	0.0009
D	0.23	0.012

FIG. 6-4   Chromatogram of banana flavor concentrate:
0.02-in.-i.d. x 500-ft SF-96(50) column, FID.

components of the column effluent and will contribute significant background to all mass spectra recorded at high column temperatures.
Usually the contributions from silicone compounds are easily recognized
and may be utilized as convenient mass scale references.

Wide ranges of component concentrations are common in mixtures of volatiles isolated from foods. Minor components may represent less than 0.1% of mixtures containing more than 100 components. Guadagni et al. (25) found that methyl thiohexanoate, which represented only 0.12% of the mass of hop oil, accounted for 4.8% of the total odor intensity. Minor or trace components of the volatile mixtures isolated from foods cannot be neglected, since they may be potent odorants. The presence of large numbers of components, containing a variety of functional groups in widely varying proportions, seriously aggravates separation problems. The probability of detecting minor components decreases as the number of components increases. The minor or trace components may be completely masked by large peaks of the major components even on the most efficient columns. In the case of open tubular columns, the maximum amount which can be applied is limited by the permissible amounts of major components.

Prefractionation is often desirable to reduce the complexity of mixtures obtained from foods and to simplify separations. When GC-MS analysis is applied, it is also necessary to consider the amounts and the dynamic range of amounts entering the spectrometer. Though it is most desirable to employ the highest efficiency columns possible in GC-MS systems, it is not usually wise to inject the complete flavor concentrate or essence. Preparative gas chromatography, followed by final separation of collected fractions on high efficiency columns, permits enrichment of minor components. Selectivity is utilized by using different liquid phases for preparative and final analysis.

After components of flavor concentrates have been identified, it is desirable to check for artifacts produced during sample separation and processing, and for contamination introduced by the procedures used. Direct head space analysis appears to be the most effective means to accomplish these goals since the minimum amount of sample handling is required. The ideal approach would be to combine the best features of all available methods, that is, to employ high efficiency columns with mass spectrometric analysis. The feasibility of direct food vapor analysis by combined open tubular GC-MS has been demonstrated by Heins et al. (26) (also see Chapter 4), but applicability of the method appeared to be limited to major components of the vapor phase. Two problems were encountered. The 0.01-in.-i.d. open tubular column was blocked by ice formed from the large excess of water vapor during subambient temperature-programmed operation. Elution of water vapor during analysis on a polar column caused interference. Flath et al. (27) overcame these problems by using larger bore columns and external traps for concentrating the vapor.

A similar technique has been used for examination of head space from ripe bananas by open tubular GC-MS analysis (24). A typical open tubular column chromatogram is shown in Fig. 6-5. Five 20-cc samples were injected into a trap consisting of a 12-in. coil of the column which was removed from the chromatograph oven and cooled in liquid nitrogen. After injection of the last sample, a 30-min period was required to establish steady flow and to elute noncondensed gases. Only injection of the last 20-cc sample is shown in Fig. 6-5. The rectangular deflection in the TIC record at the beginning of the chromatogram was caused by the large quantity of air injected. Mass spectra recorded during that period indicated separation of oxygen and nitrogen during elution from the cold trap. After the equilibration period, coolant was quickly removed. The coil was then heated with a hot air blower and returned to the chromatograph oven. Column temperature was programmed from 50 to 150° C. Although this technique is very convenient and is a very useful approach to analysis of food volatiles, it is subject to even more severe dynamic range limitations than those previously described for flavor concentrates or essences.

Conventional vapor sampling methods discriminate against low volatility components which may be present in low concentrations in the vapor phase but contribute significantly to aroma. Use of large vapor

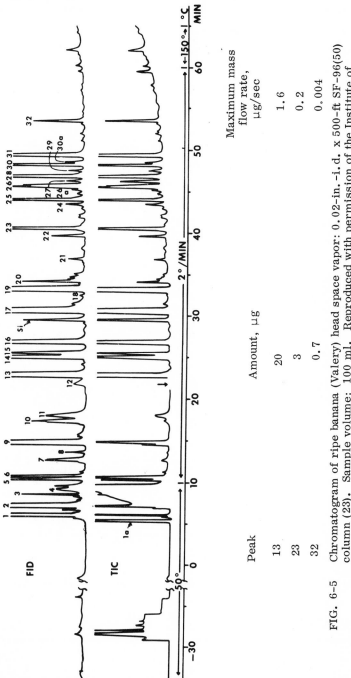

Peak	Amount, μg	Maximum mass flow rate, μg/sec
13	20	1.6
23	3	0.2
32	0.7	0.004

FIG. 6-5   Chromatogram of ripe banana (Valery) head space vapor: 0.02-in.-i.d. x 500-ft SF-96(50) column (23). Sample volume: 100 ml. Reproduced with permission of the Institute of Food Technologists.

volumes with efficient columns can minimize this problem. Under
ideal conditions, vapor analysis can be considered a special case of
the prefractionation approach recommended above. Banana flavor concentrates contain large quantities of volatile alcohols which, since
they are relatively water soluble and are therefore present in reduced
concentrations in vapor samples (28), do not interfere with identification of the small quantities of esters present. Though alcohols pre-
dominate in chromatograms of banana concentrates, vapor samples
are greatly enriched in ester components. Identities of some components of this mixture are listed in Table 6-1. In previous investigations, esters of isobutyric, valeric, and isovaleric acids were never
detected. These were identified by direct capillary column GC-MS
analysis of the head space vapor from ripe bananas (29). The range of
sample quantities present in this head space sample and the peak flow
rates measured are also indicated in Fig. 6-5. Isoamyl isovalerate,
with a maximum sample flow rate (at the column outlet) of 40 ng/sec,
produced an intense mass spectrum adequate for unambiguous identification of this compound even after passage through a separator operated at an efficiency of about 20%. Comparison of the records from
the FID and the TIC monitor indicates that no significant band spreading
occurred in the heated connecting line, carrier gas separator, or
ion source.

II. AUXILIARY TECHNIQUES

A. Catalytic Hydrogenation

All available additional information should be employed to aid
interpretation of mass spectra recorded during GC-MS analysis. Retention data are recorded simultaneously and may be used to resolve
ambiguities which remain after interpretation of spectra.

Many compounds which occur in natural flavor isolates are unsaturated; therefore, hydrogenation can be a very useful technique.
Simple on-line catalytic hydrogenators have been described (30, 31)

TABLE 6-1

Components Identified by Open Tubular Column
GC-MS Analysis of Banana Vapor
(see Fig. 6-5) (28)

Peak no.	Identity	Peak no.	Identity
1	Acetaldehyde	17	2-Pentyl acetate
2	Ethanol	18	i-Butyl i-butyrate
4	Water	19	i-Amyl acetate
6	Ethyl acetate	21	n-Amyl acetate
9	2-Pentanone	23	i-Butyl butyrate
10a	i-propyl acetate	25	n-Butyl butyrate
11	2-Pentanal	26a	i-Butyl i-valerate[a]
12	i-Amyl alcohol	26	n-Hexyl acetate
13	i-Butyl acetate	27	n-Amyl i-butyrate[a]
14	Ethyl butyrate	28	2-Pentyl butyrate
15	n-Hexanol	30a	n-Butyl n-valerate[a]
16	n-Butyl acetate	31	i-Amyl butyrate
		32	i-Amyl i-valerate[a]

[a]Not previously identified in studies of banana odor concentrates.

for use between the injection port and the column. Placement of the
hydrogenator at this point is useful for characterization of pure materials or mixtures containing a limited number of components. Difficulties arise in complex mixtures since many isomers are converted
to the same product, resulting in uncertainty in relating product peaks
with peaks in chromatograms obtained without hydrogenation. When a
GC-MS system is available, the ideal location for a microcatalytic
hydrogenator is between the chromatographic column and the mass
spectrometer. Peaks appearing on the FID or TIC chromatogram

have the retention times of the original components, thereby making correlation with ordinary chromatograms possible; but they yield spectra of the hydrogenated products. This approach has been applied to qualitative analysis of α-olefins (32). The method can also provide supplementary structural information if spectra of unreacted compounds and hydrogenated products are compared. Such comparisons are useful in an analysis of compounds containing both rings and double bonds. The empirical formulas of such compounds, determined by mass spectrometry, show the number of rings plus double bonds. Hydrogenation of double bonds, under conditions which prevent ring opening, can eliminate the ambiguity in spectrum interpretation. The possibility that rearrangements can occur must be considered.

Figure 6-6 is a schematic diagram of a microreactor used in some initial studies of catalytic on-line hydrogenation (24). The reactor, along with the connecting and valving systems, was placed in the detector oven of the inlet chromatograph and maintained at a tempera-ture of 145°C. The switching valves provide a means for bypassing the reactor when spectra of original material are acquired. The re-actor contained 10-15 mg of 1% neutral palladium catalyst on Chromosorb W (30).

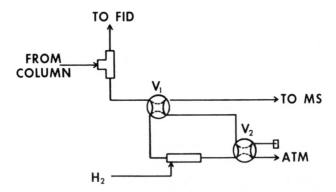

FIG. 6-6   On-line catalytic hydrogenator (24). Reproduced
with permission of the American Chemical Society.

Results of application of this microreactor to analysis of a mixture of hexen-1-ols is shown in Fig. 6-7 (24). It appears that the large excess of Chromosorb present causes slightly increased retention and tailing of the alcohol peaks during passage through the hydrogenator. The quantity of each alcohol injected was between 0.2 and 0.6 µg. The chromatograms were recorded by the TIC monitor.

Mass spectra, recorded in 1.5 sec (for the mass range m/e 20-200) during elution of each peak, with the column effluent bypassing the hydrogenator, are shown in Fig. 6-8. Spectra of peaks 5 and 6 are consistent with published spectra of trans-4- and cis-4-hexenols, respectively (33). Peak 4 (trans-2-hexenol) was not well separated from peak 5 and the spectrum shows a major contribution from the trans-4 isomer at m/e 67. Peak 3 contains cis-3-hexenol. Although peak 3 is well separated from all other components, its spectrum does not agree with published data. In the spectrum shown, the base peak (100%) is at m/e 67 and the m/e 41 peak has about 80% relative intensity. These ratios are in reverse order of those reported by Honkanen et al. (33). This discrepancy is probably due to the poor precision resulting from the fast scan rate employed. Peak 1 in the chromatogram was n-hexanol. From interpretation of the mass spectra, one can conclude only that peaks 2 through 6 contain hexenols. Though it is possible to determine double-bond position in these isomers from mass spectra recorded at slower speeds, statistical variations and other noise prevent positive identification from the 1.5-sec scans. The mixture of hexen-1-ols was hydrogenated using the microreactor, and the spectra obtained are shown in Fig. 6-9. All peaks can be identified as n-hexanol, though some still exhibit small peaks at m/e 82 due to the presence of unreacted material. After determination by mass spectrometry that all of the peaks contain hexenols which are reduced to n-hexanol, branched isomers and secondary alcohols could be excluded. Homogeneity of each peak could also be determined. Gas chromatographic retention data can then be employed to refine the characterization. This technique was

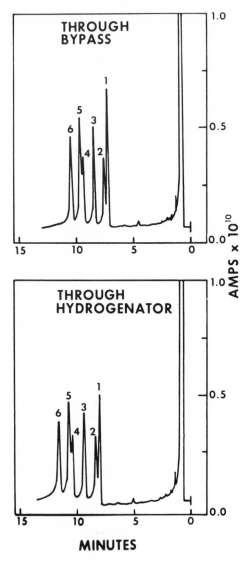

FIG. 6-7   Chromatograms of isomeric hexenols: 0.02-in.-i.d. x
50-ft Carbowax 1540 SCOT column, TIC detection.
Peak identities: 1, n-hexan-1-ol; 2, trans-hex-3-en-
1-ol; 3, cis-hex-3-en-1-ol; 4, trans-hex-2-en-1-ol;
5, trans-hex-4-en-1-ol; 6, cis-hex-4-en-1-ol (24).
Reproduced with permission of the American Chemical Society.

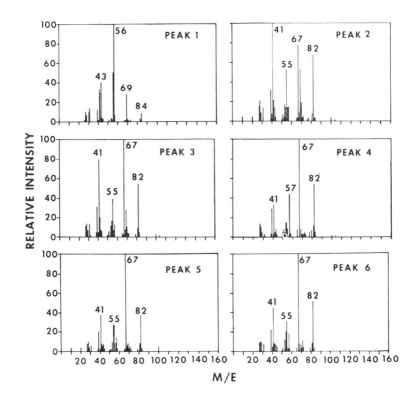

FIG. 6-8   Mass spectra of isomeric hexenols. Peak numbers
            correspond to those of Fig. 6-7 (24). Reproduced
            with permission of the American Chemical Society.

used to confirm identification of cis- and trans-4-hexen-1-ol in
hydrolyzed banana concentrate (34).

The major limitation of the on-line hydrogenator is that temperatures must be high enough to prevent sample condensation but low
enough to prevent hydrogenolysis. This is relatively easy to do with
compounds of high volatility, but it is impossible with some materials.
Microbatch hydrogenation is generally applicable to relatively pure
fractions collected from GC columns, and is most conveniently applied
to samples of low volatility. Figure 6-10 is a chromatogram of a
phenolic fraction from wood smoke condensate (35). Two components,

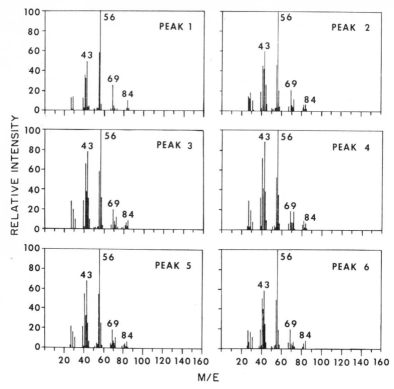

FIG. 6-9   Mass spectra of hydrogenated hexenols. Peak numbers correspond to those of Fig. 6-7 (24). Reproduced with permission of the American Chemical
Society.

fractions 18 and 20, with molecular weight 194 and identical mass
spectra were detected during GC-MS analysis.

Figure 6-11A is the mass spectrum of fraction 18, and Fig. 6-11B
is the spectrum of fraction 20. Both were recorded during elution
from a 1/8 in. x 6 ft SE-30 packed column. A third component
(fraction 17) also exhibited a similar spectrum with a molecular ion at
m/e 194, but the peak at m/e 167 (M-27) was significantly more
intense (25%) than in the spectra of fractions 18 and 20. The three
fractions were hydrogenated utilizing Adam's platinum oxide catalyst

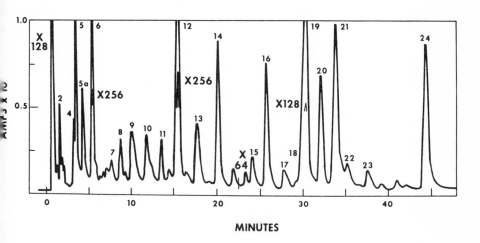

FIG. 6-10   Chromatogram of phenolic fraction of hardwood
             smoke: 1/8-in.-o.d. x 6-ft 3% SE-30 Column,
             FID (35). Reproduced with permission of the
             American Chemical Society.

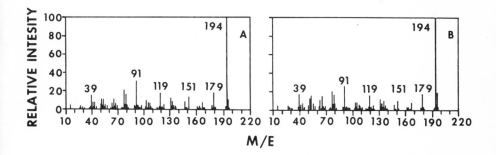

FIG. 6-11   Mass spectra of isomeric phenols from hardwood
             smoke. A, fraction 18, B, fraction 20 (35).
             Reproduced with permission of the American
             Chemical Society.

(0.5 mg) in ethanol under hydrogen for 1 hr. Quantities of unknown
fractions were between 10 and 50 µg. The reaction mixtures were
analyzed directly by GC-MS and, in all three cases, the spectra of
the hydrogenated products were identical with the spectrum of
4-propyl-2,6-dimethoxyphenol. Typical spectra are shown in Fig.
6-12. Figure 6-12A is the spectrum recorded during GC-MS analysis
of the hydrogenation product of fraction 20. The spectrum of the
product and retention data on SE-30 and Carbowax 20M-TPA columns
were identical with those recorded for 4-propyl-2,6-dimethoxyphenol.
Infrared spectra confirmed assignment of the substitution pattern.
These results indicated that the substituent in the <u>para</u> position
contained three carbon atoms, but the position of the double bond
could not be determined directly from the mass spectral data.

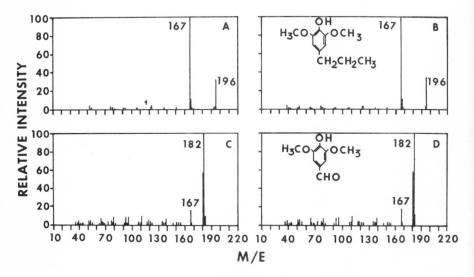

FIG. 6-12   A, Mass spectrum of hydrogenation product of frac-
             tion 20; B, Mass spectrum of authentic 4-propyl-2,6-
             dimethoxyphenol; C, Mass spectrum of ozonolysis
             product of fraction 20; D, Mass spectrum of authentic
             syringaldehyde (35). Reproduced with permission of
             the American Chemical Society.

B. Ozonolysis

Ozonolysis is a useful method for cleavage of double bonds. The feasibility of this technique for determination of double-bond position in a wide range of organic compounds on a microgram scale has been demonstrated by Beroza and Bierl (36). They employed a simple generator in which ozone is produced by a high voltage discharge in oxygen. Relatively low concentrations of ozone are produced in this apparatus (approximately 0.2 mole %) (37). Triphenylphosphine was recommended for reduction of the ozonides to aldehydes which could then be analyzed by gas chromatography (36).

This procedure was applied to fractions 17, 18, and 20 collected from the phenolic fraction of hardwood sawdust smoke. Quantities of unknown material between 10 and 50 μg and reaction times between 10 and 50 sec were employed. Fractions 18 and 20 yielded syringaldehyde. The mass spectrum of the aldehyde produced from ozonolysis of fraction 20 and the spectrum of an authentic sample of syringaldehyde are shown in Figs. 6-12C and 6-12D, respectively. Reduction of the ozonides with triphenylphosphine was not feasible since the reagent and its oxidation product yielded peaks which interfered with the chromatographic analysis of the aldehydes. Reduction using tetracyanoethylene (TCNE) produced no chromatographic or mass spectrometric interference. Ozonolysis did not proceed to any significant degree with fraction 17 under the same conditions. Apparently, the low concentration of ozone produced by the high voltage discharge results in reaction rates for some materials that are too low to be of analytical value. Thirty-minute reaction times have been reported to be necessary for some compounds (37). Presumably, fraction 17 could have been ozonized using a generator capable of producing higher ozone concentrations. Apparently, double bonds allylic to aromatic rings are difficult to ozonize under the conditions employed. In experiments with pure reference materials, eugenol (2-methoxy-4-allylphenol) could not be ozonized, while isoeugenol (2-methoxy-4-propenylphenol) reacted rapidly. On the basis of results obtained by hydrogenation and ozonolysis, fraction 18 was identified as 2,6-dimethoxy-4-<u>cis</u>-propenylphenol. The assignment of the <u>cis</u> and <u>trans</u> configurations was made on the basis of analogy of their retention behavior on SE-30 and Carbowax 20M columns to data recorded on the same columns for <u>cis</u> and <u>trans</u> isoeugenol. Microozonolysis can be used routinely for location of double bonds in relatively simple unsaturated compounds. Isolation of products is not necessary. Aldehydes produced by reductive cleavage of the ozonides can be analyzed by direct GC-MS examination. Sample quantities between 10 and 50 μg are easily handled, and extension to smaller quantities is feasible.

C. Other Methods for Location of Double Bonds

Determination of double-bond position in complex molecules by mass spectrometry represents an important and difficult problem. Double-bond migration usually follows ionization by electron impact, so that double-bond position isomers usually give identical spectra. In some cases small differences in relative intensities can be observed in spectra of isomers, but these differences may not be significant when fast scanning is used and precision of relative intensity measurement is poor. Only acyclic compounds with terminal double bonds, or compounds in which the double-bond position is stabilized by a functional group, produce spectral patterns sufficiently unequivocal to permit their unambiguous identification. An obvious solution would be to employ catalytic deuteration with subsequent examination of the mass spectra to determine the location of the deuterium label. Unfortunately, this leads to extensive isotope incorporation and scrambling of deuterium and hydrogen (38). Reduction with deuterohydrazine results in partial H-D exchange and therefore is not adequate (39).

McCloskey and McClelland (40) examined the mass spectra of O-isopropylidine derivatives formed from the dihydroxy compounds produced by oxidation of monounsaturated fatty esters. OsO_4 was used for preparation of the diols. Reagent enriched in O^{18} may be used to produce labeled derivatives, which could be useful in examination of unsaturated compounds which also have oxygen-containing functional groups.

Niehaus and Ryhage (41) employed a similar approach, oxidizing polyunsaturated fatty acids to the corresponding polyhydroxy compounds. After methylation of the polyhydroxy acids, the polymethoxy methyl esters were analyzed directly by GC-MS. The results indicate that this approach will be useful for locating unsaturation in compounds containing multiple double bonds. For those compounds in which a functional group containing oxygen is present, it also appears feasible to employ O^{18} labeling.

D. Other Synthetic and Degradative Reactions

The extent to which identification of an unknown compound is pursued depends on the certainty of structure that is required in the individual problem (see Chapter 3, Fig. 3-1). The only way to make a truly positive identification of an unknown compound isolated from a complex mixture is by comparison of all relevant physical properties with those of known pure materials. All possible isomers should be included among the reference compounds. When only microgram or submicrogram quantities are available, the only reliable physical properties that can be obtained are gas chromatographic and mass spectral data. A conservative chemist is reluctant to rely solely upon interpretation of these data for identification. Even when spectra have been reported in the literature, these may have been recorded under conditions not comparable to those used in GC-MS. Mass spectra, unlike other spectral data, are not physical properties of molecules but are subject to variability caused by experimental conditions. Therefore, it is highly desirable to record the mass spectra of the reference materials on the same instrument used for examination of unknown samples. Spectra in the literature may be those of compounds of lesser purity than can be obtained by GC-MS.

Synthesis of all possible reference materials is often neither feasible nor justifiable because of the large number of unknown compounds

which need to be identified in most flavor investigations. The examples given above from studies of the phenolic fraction of wood smoke
illustrate that unambiguous identification of unknown compounds can
be made even though pure synthetic reference materials are not available. Conversion of the unknown material by known synthetic or
degradative pathways to compounds for which reference materials are
available is adequate for some identification purposes. The greater
the variety of reactions applied to each unknown, the more positive
will be the degree of identification. Many synthetic and degradative
reactions, other than those discussed above, are applicable to microgram quantities of the flavor components which can be collected in
relatively pure form from GC columns. Only a small number have
been applied because isolation of reaction products is difficult or impossible at this level. However, when GC-MS is available, reaction
mixtures can be analyzed. Methylation, acylation, oxidation, and reduction have been used for characterization of a variety of compounds.
Most of these reactions can be scaled down to the microgram level
and would be particularly appropriate in investigations of the components of flavors. GC-MS analysis of appropriate derivatives and
reaction products can provide a greater degree of certainty of identification than GC-MS of the original compounds.

III. ACQUISITION AND PROCESSING OF MASS SPECTRAL DATA

A. The Nature of Mass Spectral Data

The normal mode of presentation of mass spectral data differs significantly from that for data obtained from NMR, IR, and UV spectrometers. The charts have no precalibrated mass scale. Ion current is
usually recorded by multichannel galvanometer traces at different
sensitivity ratios. The chemist must select the largest signal that is
on scale, multiply it by the correct sensitivity factor, and compute the
intensity of each peak relative to that of the most intense (base, or

100%) peak. Determination of the mass corresponding to each peak also presents a problem since the mass scale may not be reproducible, especially when fast scanning magnetic deflection mass spectrometers are employed. Magnet hysteresis can cause small expansions or contractions of the mass-time scale, which make unambiguous determination of mass from measurements of distance on the chart impossible. However, in low resolution mass spectrometry, determination of mass must be unambiguous. An error of one mass unit can lead to totally incorrect interpretation. The problem of establishing the correct scale is more severe in those instruments which employ a nonlinear scan function. Usually the sensitivity and dynamic range of mass spectrometers are such that a peak occurs at almost every mass, from low intensity sample fragment ions and from background in the instrument. When peaks are present at every mass, simple counting can establish the mass scale. When fast scanning is employed, the dynamic range of the instrument is reduced to the extent that normal background and small fragment ion peaks may not be observed at every mass and interpolation may be necessary. It may be desirable to admit a small quantity of mass reference compound, usually a perfluorinated hydrocarbon mixture which does not interfere with sample peaks, either continuously or intermittently (8) during fast scan operation.

In spite of the difficulties involved, determination of mass in low resolution mass spectrometry is simple compared with mass measurement in high resolution mass spectrometry. In the latter case, mass determination to an accuracy of 1-3 millimass units is usually required. A mass reference compound is always introduced with the sample, and mass differences or ratios are measured.

When fast scanning is employed, peak widths are very small. At a 5-sec/decade exponential scan at resolving power 500, the width of a peak (the time during which the peak is passing across the collector slit) is about 4 msec. At resolving power 10,000 and 10-sec/decade

scan rate, peak widths of approximately 400 μsec can be expected. In
order to make the low resolution fast scan spectra readable on the
chart paper, relatively high chart paper speeds are required. Eight
inches per second is a typical speed, so recording a 5-sec scan requires 40 in. of chart paper. Recording of complete high resolution
spectra on a chart is not feasible.

The time required for manual background subtraction and computation of a normalized low resolution spectrum varies depending on the
number of peaks present, but it is at least 15-20 min/spectrum. If
we assume that an average of 200 spectra/week are processed, then
3000 man hours/year must be spent in data-processing activities.
Occasional errors are unavoidable when such quantities of data are
processed by people. Errors are difficult to detect and correct
because of the large volume of raw data which must be reexamined.
A GC-MS system is capable of producing far more data than can be
handled by presently available noncomputer methods.

B. Automatic Data Acquisition and Processing

The need for computer processing is more obvious in the case of
high resolution mass spectral data. In these spectra hundreds of
peaks may be present, and maximum structural information is obtained
only if all are measured with accuracies requiring six or seven significant figures. Data may then be presented as an "element map" (42)
in which the spectrum is displayed as an array according to the elemental composition of the fragment ions, rather than simply as a list of
six-digit numbers. "Element maps" provide graphic representation
of the structural relationships within the molecule. This method of
data presentation was originally developed for use with photographic
plates on which all peaks are recorded simultaneously, and data may
be processed subsequently.

Digital systems have been developed for acquisition of high resolution data from instruments which employ electron multiplier detection and FM tape recording (43). On-line computers have been

employed for direct acquisition and processing (44).   Variations on
the basic "element map" presentation have been suggested (44, 45).
An on-line data acquisition system which prepares a digital magnetic
tape for subsequent computer processing has been described (46).
This system provides very efficient data acquisition without requiring
continuous access to a complete computer system.   However, in view
of the recent development of small, low cost, general-purpose computers with suitable input and output devices, specialized data acquisition systems now appear less desirable.

Because of the obvious requirement for a computer to process high
resolution mass spectral data, few laboratories would consider
purchasing a high resolution mass spectrometer without some provision for data-processing facilities.   Though it is possible to operate a
low resolution fast scanning mass spectrometer with manual processing of all data, the desirability of an on-line processing system for
such instruments should be considered.   Recent developments in integrated circuit technology have resulted in significant reductions in cost
of small general-purpose computer systems.   Basic systems can be
purchased for less than $10,000.   Complete systems, including input
and output devices, analog-to-digital converters, and mass memory
systems (magnetic tape, drums, disks), can be expected to cost
between $30,000 and $50,000.

A special-purpose data acquisition system, which prepares a digital magnetic tape output suitable for computer processing, has been
used in GC-MS (47).   This system was used with a low resolution magnetic deflection mass spectrometer operated under fast scan conditions.   In order to attain a relatively reproducible mass scale, it was
necessary to operate in a repetitive scanning mode to establish a
stable cycle of the magnetic field.   Small differences in mass scale
between successive scans were corrected by a computer program
which used the output of the digitizing system as its input.   Spectra
can be recorded continually during a GC-MS experiment in order to
extract the maximum amount of information obtainable.   A reel of

digital tape can contain 300 spectra, corresponding to a chromatogram 20 min long. Ten to 20 digital samples are taken on each mass peak at the analog-to-digital conversion rate of 3000/sec. One disadvantage of such a system is that data are recorded on the output tape even when no spectral peaks are present; i.e., the output tape contains a large proportion of zero data points. The cost of this special-purpose data acquisition system approaches that of a small general-purpose computer including appropriate input and output devices. It was estimated that the processing of one complete spectrum takes 20 sec on an IBM 7094 computer. This time saving, compared to manual processing, can be translated into a substantial saving in data-processing costs. In addition, once the spectrum is in computer-compatible form, other operations may be performed by the computer. These include subtraction of background, subtraction of spectra for determination of unresolved components, and comparison with reference spectra for tentative identification of unknown compounds.

Hites and Biemann (48) described an on-line computer system that is well suited to GC-MS operation. A medium sized computer (IBM-1800) was employed. Repetitive scanning of the magnetic field was used and spectra were recorded continually during GC-MS analysis. Digitized spectra were processed during acquisition and stored on magnetic disks. Since all spectra were stored, it was possible to construct a novel "total ionization" chromatogram from the mass spectral data. This chromatogram was reconstructed by adding all mass peaks from each spectrum and plotting the total as a single point. The method is analogous to that employed in integrating TIC monitors used in time-of-flight and quadrupole mass spectrometers except that chromatograms can be reconstructed at any time after recording. After examination of a "total ionization" chromatogram, the spectrum at any point can be plotted for individual examination. Continuous recording relieves the operator of responsibility for determining the best time to scan a spectrum.

A means for establishing an unambiguous mass scale would be valuable both for oscillographically recorded spectra and for spectra recorded by on-line computers. Jansson et al. (49) described a high speed mass marker system, employing a Hall-effect transducer for measurement of magnetic field. It was used as part of a digital data acquisition system which prepared a magnetic tape for subsequent use as computer input. The output of the digital circuitry may be displayed on the instrument control panel, recorded with the spectrum on a galvanometer trace or analog magnetic tape, or used as input for an on-line computer. Such a system can be constructed to operate with most magnetically scanning mass spectrometers.

Provision of a mass identification signal from magnetic deflection spectrometers would eliminate the necessity for mass determination by interpolation and extrapolation from known reference peaks. The resultant savings in computer time and capacity could make application of a small general-purpose computer to complete on-line data acquisition, storage, and processing of GC-MS data feasible, even when hundreds of spectra are scanned during a long GC analysis. Instead of storing all points accumulated during analog-to-digital conversion, a threshold value can be set by the computer program or by an external device and digitization can be initiated only when the threshold is exceeded. At the time digitizing begins, the output of the mass measurement circuit can be measured to provide a single unambiguous value for later conversion to mass by the computer. All of the digital samples converted during the time that the ion current is above threshold can be added together to provide an integrated value of ion current for each peak. It is actually easier and faster for a computer to integrate in this manner than to determine the peak height from the noisy signals which can be expected during fast scan operation. The integration process provides a smoothing effect which increases pattern precision, and it provides intensity data that are independent of scan function. Only the integral and the field value, or

mass equivalent to the field, need be stored. Storage capacity required in the computer or in peripheral mass storage devices can be
reduced. The dynamic range of most mass spectrometers can be
adequately handled by the 12-bit (4096:1) resolution of available
analog-to-digital converters provided with small computer systems.

A computer employed in this manner would probably be connected
in parallel with the conventional output system used during repetitive
rapid scanning. The operator could still monitor the spectrum,
in analog form, at all times on the oscilloscope screen. The major
advantage is that no data are lost during a prolonged GC-MS analysis.
All spectra are recorded in digital form and can be recovered later
for examination. The ability to recover data after completion of the
experiment is an important consideration. Many chemists have
wished that they had recorded a spectrum a few seconds earlier or
later to determine whether certain peaks were due to a major component of the chromatographic peak or to some unresolved component,
background, or previous peak.

A complete "closed loop" control system for time-of-flight and
quadrupole mass analyzers was described by Reynolds et al. (50).
The computer was used to generate the scan function, a voltage rather
than a magnetic field value, so the system was not subject to hysteresis
effects. The computer, a LINC with a 2000-word memory and a tape-
operating system, generated the scan, acquired spectral data, did com-
putations, and prepared output. These operations required a comprehensive programming (software) system. The mass scale generating
function was calibrated from time to time by introduction of a reference fluorocarbon compound. The mass range of the system, m/e
1 to 256, was limited by the hardware available. It is likely that
newer computers could be employed in similar systems to provide
the same features without these limitations.

C. Minimum Cost Approach

It should be clear from the previous discussion that on-line computers or special-purpose digital data acquisition systems (which are
really special-purpose digital computers) can be extremely useful but
at the same time very expensive. Even when a small laboratory-type
computer is employed, the cost of the data-processing system may be
equal to or may exceed that of the mass spectrometer used for generating the data. Even though long-term labor savings may justify
such an expenditure, it is often difficult to acquire adequate funds for
purchase of both the mass spectrometer and the data-processing system at the same time. A compromise system is required which would
eliminate tedium and labor involved in processing the large numbers
of mass spectra which can be produced by modern fast scanning instruments.

Analog FM tape recording appears to be a suitable solution for recording repetitive spectra for subsequent processing by a remotely
located computer or for manual processing. FM tape recorders (see
Chapter 5) are now available at relatively low cost. The magnetic
tape employed is inexpensive and may be reused after the data contained on it are analyzed or discarded. Spectra can be monitored via
an oscilloscope screen connected in parallel with the FM tape recorder.
Total ion current or auxiliary gas chromatographic detector output can
also be recorded on one track of the magnetic tape. The output of a
mass marker circuit and a timing signal can be recorded on separate
tape channels. A conventional oscillograph permits recording of
spectra on paper if desired.

Jansson et al. (49) described a system in which the number of required tape channels is reduced by recording the mass spectrometer
ion current output on a single channel after conditioning by a logarith-
mic amplifier. The data, recorded in analog form, may be checked

by observation on the oscilloscope screen prior to transporting the
tape to the computation facility.

D. Computer Identification of Unknown Compounds

Once mass spectra have been processed into tables of relative
intensity vs. mass number, the next logical step is to employ a computer for comparison of unknown spectra with spectra of reference
compounds. Assuming that all previously recorded spectra could be
stored on magnetic tape files for examination by a computer, it is
still necessary to optimize searching routines to provide maximum
information in minimum computer time. Comparison of all peaks in
the unknown spectrum with all peaks of spectra in the reference collection results in minimum ambiguity of matching (51) but wastes
computer time. Considerable time is saved by selection of a fixed
number (five or six) of the most intense or most characteristic peaks
for comparison with reference data (51, 52).

A reasonable compromise has been suggested by Hites and Biemann
(53), who employed an "abbreviated" mass spectrum for computer
comparisons with similarly treated spectra of reference compounds.
The abbreviated spectrum eliminates many of the minor peaks, which
often provide little structural information. Selecting the two largest
peaks from each 14-mass-unit interval results in a greatly reduced
amount of data but preserves most of the structural information.

Spectra recorded on different instruments may exhibit systematic
differences related to variations in ionizing energy, source temperature, scan function, and other experimental conditions. A means for
comparison of an unknown spectrum with reference spectra recorded
on a variety of instruments should, ideally, be able to compensate
for these systematic differences. The chi-square test, applied to a
multiple regression analysis performed by computer, was shown to
be useful for distinguishing differences due to fragmentation pattern
from those caused by systematic mass discrimination (53). Crawford

and Morrison (54) evaluated statistical procedures for testing similarity of spectra in identification of unknown compounds by comparison with a catalog of known spectra. They concluded that use of the six strongest peaks resulted in the fastest comparisons (1.7 sec/compound) and was adequate for identification of most compounds examined.

Searches of reference spectra files for comparison of unknown with known spectra represent the simplest means of identification. Often it is not necessary to search the entire catalog of known compounds, since restrictions can be placed on molecular weight or sample type based on other information or on the mass spectral data. These limitations can be included in the computer program or imposed by the operator if a small computer is employed for searching. The methods described are suitable for matching spectra of unknown compounds with those that have been previously recorded and are cataloged in reference files. Difficulties arise in attempting to determine the structure of compounds for which spectra have not been previously recorded or published. For these compounds, interpretation of the spectrum is necessary. Techniques described previously may be useful in converting the unknown to a compound for which the spectrum is documented. Final positive identification still requires recording of mass spectra and other relevant data, particularly GC retention data, of known synthetic reference materials.

E. Recent Developments

Computer methods for identification of compounds for which reference mass spectral data are not cataloged would be welcomed. The data provided by high resolution mass spectrometry are sufficiently detailed in terms of elemental composition that computer methods for their interpretation have been investigated. Some successes have been reported in computer-aided interpretation of the high resolution mass spectrum of a sulfur-containing component from an enzymatic hydrolysate of yeast sRNA (55), and interpretation of high resolution mass spectra of relatively complex peptides (56-58). Less progress

has been made in computer-aided interpretation of low resolution
mass spectra. Crawford and Morrison (59) reported an approach to
determination of molecular class of unknown compounds, based on low
resolution mass spectral data, and Jurs et al. (60) investigated the
use of a computer for molecular formula determination from low
resolution mass spectrometric data. Computer methods for searching
reference catalogs for identification of compounds which are present
in these files, are well-developed, practical, time-saving systems.
The use of digital computers for actual interpretation of mass spectra
to yield structural information for compounds not present in reference
file collections is still in an experimental stage. Rapid development
in this area may be expected.

Use of computers for acquisition, computation, and analysis of
spectra, and for searching reference spectra files, represents a
time-saving device. These are jobs that can be done manually if one
is willing to tolerate the inefficiency. The major contribution of these
systems is that simple compounds, present in complex multicomponent
mixtures, can be identified very rapidly. The chemist can then devote
his attention to more complex interesting components. This latter
group represents the maximum challenge and will require more than
mass spectra for characterization.

IV. SUMMARY

Combined GC-MS analysis provides the flavor chemist with a means
for rapid analysis and identification of large numbers of components
present in the mixtures that he isolates. It is, therefore, perfectly
suited to the types of samples encountered in most investigations of
volatile flavor components. By combining the GC-MS system with
computer analysis of data and searching of reference files, it is possible to identify rapidly large numbers of simple compounds which
occur in these complex mixtures. Elimination of these compounds

frees the investigator to pursue more difficult structure problems.

A GC-MS system is capable of recording spectra on full-range flavor concentrates, head gas samples, and essences, separated on high efficiency GC columns; however, prefractionation by preparative GC, thin layer chromatography, or column chromatography is usually desirable to simplify these mixtures prior to GC-MS analysis. Prefractionation results in reduction of the dynamic range of sample quantities introduced into the mass spectrometer, thereby permitting adjustment of operating parameters to produce the highest quality spectrum of each component. Prefractionation also provides the investigator with the opportunity to optimize separation on high resolution GC columns.

A variety of methods is available for interfacing a gas chromatograph with a mass spectrometer. Variations in operating parameters with each type of interface are possible. It is usually necessary to optimize each system for the particular sample, column flow rate, and mass spectrometric pumping system employed. No single GC-MS interface system can be considered better than all others for all purposes, but the membrane separator offers a particular advantage in flavor chemistry. The excess effluent, vented to the atmosphere, can be collected for further examination by other physical methods or subjected to odor evaluation during the GC-MS analysis.

GC-MS is a generally applicable system for analysis of reaction mixtures without the need for prior separation. Microdegradation and synthetic methods can be employed. Further development in this area can aid in the identification of compounds for which spectra are not available and can provide confirmation of identifications.

Maximum utilization of the data available under optimum conditions from a GC-MS analysis requires continuous recording, under repetitive scan conditions. The amounts of data generated by repetitive scanning, particularly during long gas chromatographic analyses, require use of either an analog FM tape-recording system or an online computer for acquisition, computation, and storage of spectral data. Few laboratories that are not engaged in active development of mass spectrometric methods and instrumentation employ advanced data-recording and computation systems. Other laboratories, particularly flavor research laboratories, should consider the economics of manual data processing and examine the alternative modes of recording and processing these data.

In this discussion, emphasis has been placed on identification of volatile food constituents of possible significance in the aroma of foods. Mass spectrometry is applicable to analysis of relatively nonvolatile components, or volatile derivatives prepared from these components. Many of the compounds which contribute to taste and to astringency can be examined by mass spectrometry for structural characterization.

The availability of GC-MS systems has made the flavor chemist's job easier. It is now possible to identify large numbers of components far more efficiently. However, relationships between composition and sensory quality still remain the major problems of the flavor chemist.

REFERENCES

1. A. E. Banner, 13th Ann. Conf. Mass Spectrometry and Allied Topics, St. Louis, Mo., 1965, p. 193.

2. A. J. Campbell and J. S. Halliday, 13th Ann. Conf. Mass Spectrometry and Allied Topics, St. Louis, Mo., 1965, p. 200.

3. J. H. Beynon, Advances in Mass Spectrometry (E. Kendrick, ed.), Vol. 4, The Institute of Petroleum Press, London, 1968, p. 123.

4. K. R. Jennings, J. Chem. Phys., 43, 4176 (1965).

5. M. Barber, W. A. Wolstenholme, and K. R. Jennings, Nature, 214, 664 (1967).

6. A. H. Struck and H. W. Major, Jr., 17th Ann. Conf. Mass Spectrometry and Allied Topics, Dallas, Texas 1969, p. 102.

7. T. W. Shannon, T. E. Mead, C. G. Warner, and F. W. McLafferty, Anal. Chem., 39, 1749 (1967).

8. F. A. J. M. Leemans and J. A. McCloskey, J. Am. Oil Chem. Soc., 44, 11 (1967).

9. W. H. McFadden, R. Teranishi, D. R. Black, and J. C. Day, J. Food Sci., 28, 316 (1963).

10. W. H. McFadden and R. Teranishi, Nature, 200, 329 (1963).

11. R. Teranishi, R. G. Buttery, W. H. McFadden, T. R. Mon, and J. Wasserman, Anal. Chem., 36, 1509 (1964).

12. R. S. Gohlke, Anal. Chem., 31, 535 (1959).

13. J. T. Watson and K. Biemann, Anal. Chem., 37, 845 (1965).

14. M. A. Grayson and C. J. Wolf, Anal. Chem., 39, 1438 (1967).

15. M. C. ten Noever de Brauw and C. Brunnee, Z. Anal. Chem., 229, 321 (1967).

16. W. D. MacLeod, Jr., and B. Nagy, Anal. Chem., 40, 841 (1968).

17. M. Blumer, Anal. Chem., 40, 1590 (1968).

18. R. Ryhage, Anal. Chem., 36, 759 (1964).

19. R. Ryhage, Arkiv Kemi, 26, 305 (1967).

20. P. M. Llewellyn and D. P. Littlejohn, presented at Pittsburgh Conf. Anal. Chem. Appl. Spectry., February, 1966.

21. D. R. Black, R. A. Flath, and R. Teranishi, J. Chromatog. Sci., 7, 284 (1969).

22. C. Brunnee, L. Jenckel, and K. Kronenberger, Z. Anal. Chem., 197, 42 (1963).

23. P. Issenberg, Food Technol., 23, 1435 (1969).

24. P. Issenberg, A. Kobayashi, and T. J. Mysliwy, J. Agr. Food Chem., 17, 1377 (1969).

25. D. G. Guadagni, R. G. Buttery, and J. Harris, J. Sci. Food
     Agr., 17, 142 (1966).

26. J. Th. Heins, H. Maarse, M. C. ten Noever de Brauw, and
     C. Weurmann, J. Gas Chromatog., 4, 395 (1966).

27. R. A. Flath, R. R. Forrey, and R. Teranishi, J. Food Sci.,
     34, 382 (1969).

28. M. J. Myers, Ph.D. Thesis, Mass. Inst. Technol., 1968.

29. S. Quast and T. J. Mysliwy, Mass. Inst. Technol., unpublished
     results, 1969.

30. M. Beroza and R. Sarmiento, Anal. Chem., 38, 1042 (1966).

31. T. L. Mounts and H. J. Dutton, J. Am. Oil Chem. Soc., 44,
     67 (1967).

32. R. M. Teeter, C. F. Spencer, J. W. Green, and L. H. Smithson,
     J. Am. Oil Chem. Soc., 43, 82 (1966).

33. E. Honkanen, T. Moisio, M. Ohno, and A. Hatanaka, Acta Chem.
     Scand., 17, 15 (1963).

34. E. L. Wick, T. Yamanishi, A. Kobayashi, S. Valenzuela, and
     P. Issenberg, J. Agr. Food Chem., 17, 751 (1969).

35. A. O. Lustre and P. Issenberg, J. Agr. Food Chem., 17, 1387
     (1969).

36. M. Beroza and P. A. Bierl, Anal. Chem., 39, 1131 (1969).

37. Supelco, Inc., Chromatography-Lipids, III, No. 2, May, 1969.

38. N. Dinh-Nguyen and R. Ryhage, J. Res. Inst. Catalysis
     Hokkaido Univ., 8, 73 (1960).

39. N. Dinh-Nguyen, R. Ryhage, S. Stallberg-Stenhagen, and
     E. Stenhagen, Arkiv Kemi, 18, 393 (1961).

40. J. A. McCloskey and M. J. McClelland, J. Am. Chem. Soc.,
     87, 5090 (1965).

41. W. G. Niehaus and R. Ryhage, Tetrahedron Letters, 49, 5021
     (1967).

42. K. Biemann, P. Bommer, D. M. Desiderio, and W. J.
     McMurray, in Advances in Mass Spectrometry (W. L. Mead, ed.),
     Vol. 3, The Institute of Petroleum, London, 1966, p. 639.

43. W. J. McMurray, B. N. Greene, and S. R. Lipsky, Anal. Chem.,
    38, 1194 (1966).

44. A. L. Burlingame, D. H. Smith, and R. W. Olsen, Anal. Chem.,
    40, 13 (1968).

45. R. Venkataraghavan, R. D. Board, R. Klimowski, J. W. Amy,
    and F. W. McLafferty, in Advances in Mass Spectrometry
    (E. Kendrick, ed.), Vol. 4, The Institute of Petroleum, London,
    1968, p. 65.

46. C. Merritt, Jr., P. Issenberg, and M. L. Bazinet, in Advances
    in Mass Spectrometry (E. Kendrick, ed.), Vol. 4, The Institute
    of Petroleum, London, 1968, p. 55.

47. R. A. Hites and K. Biemann, Anal. Chem., 39, 965 (1967).

48. R. A. Hites and K. Biemann, Anal. Chem., 40, 1217 (1968).

49. P. A. Jansson, S. Melkersson, R. Ryhage, and S. Wikstrom,
    16th Ann. Conf. Mass Spectrometry and Allied Topics,
    Pittsburgh, 1968, p. 306.

50. W. E. Reynolds, J. C. Bridges, R. B. Tucker, and T. B.
    Coburn, 16th Ann. Conf. Mass Spectrometry and Allied Topics,
    Pittsburgh, 1968, p. 77.

51. B. Petterson and R. Ryhage, Arkiv Kemi, 26, 293 (1967).

52. S. Abrahamsson, Science Tools (LKB Instrument Co), 14, 29
    (1967).

53. R. A. Hites and K. Biemann in Advances in Mass Spectrometry
    (E. Kendrick, ed.), Vol. 4, The Institute of Petroleum, London,
    1968, p. 37.

54. L. R. Crawford and J. D. Morrison, Anal. Chem., 40, 1464
    (1968).

55. K. Biemann in Advances in Mass Spectrometry (E. Kendrick,
    ed.), Vol. 4, The Institute of Petroleum, London, 1968,
    p. 139.

56. M. Senn and F. W. McLafferty, Biochem. Biophys. Res.
    Commun., 23, 4 (1966).

57. K. Biemann, C. Cone, and B. R. Webster, J. Am. Chem. Soc.,
    88, 2597 (1966).

58.   M. Barber, P. Powers, M. J. Wallington, and W. A. Wolsten-
      holme, Nature, 212, 5064 (1966).

59.   L. R. Crawford and J. D. Morrison, Anal. Chem., 40, 1469
      (1968).

60.   P. C. Jurs, B. R. Kowalski, and T. Isenhour, Anal. Chem.,
      41, 21 (1969).

Chapter 7

OTHER IDENTIFICATION METHODS:
TRAPPING, NMR, IR, RAMAN

I. INTRODUCTION

Although gas chromatography-mass spectrometry (GC-MS) is the
most useful identification method in flavor research, complete structure often cannot be elucidated with MS data alone. Other data must
be consulted. Amounts needed for other spectral methods are larger;
therefore, all available techniques must be used to decrease sample
size requirements. Techniques for manipulation of small samples
without loss or contamination require attention.

II. GC TRAPPING

For obtaining most spectrometric data, samples must be trapped and transferred to cells specific to each spectrometer. Many traps and methods have been described. The fundamental problem is to quantitatively condense purified material from the gaseous state into manageable form and to be able to recover it. To prevent aerosol, or fog formation, dew points must be attained without excessive nucleation. If aerosols are formed, percentages trapped are very small unless centrifugal (1) or electrostatic (2-5) traps are used. These were designed for large-scale preparative work and are not suitable for routine work with milligrams. The simplest and most commonly used method is to reach the dew point gradually with a thermal gradient (6, 7).

A. Collectors

For very small samples, the most common device is a glass melting point capillary, Teflon, or polyethylene tube (6, 8-11) inserted into the heated GC exit. Because of the small mass, the tubing is quickly heated by the GC outlet. Gradient cooling is quickly established, and condensation occurs with minimum fog formation. If the tubing is not heated rapidly enough when inserted, it can be preheated on a hot plate or with a hot tip of a soldering gun. Of course, such trapping techniques do not apply if the dew points are lower than room temperature. If cooling is necessary, care must be taken to minimize condensation of atmospheric moisture in the trap.

In some cases, a little piece of cotton can be wrapped around the capillary, and it can be wetted with a low boiling solvent, such as ethyl ether or chloroform. Cooling by evaporation is sometimes sufficient to provide good trapping yields. In some cases, dry ice must be used. Liquid nitrogen should be avoided as a coolant because oxygen can be condensed. Violent explosions can occur with organic materials in the presence of liquid oxygen.

For efficient trapping of low boilers, it is helpful to insert a fine
wire coil inside glass capillaries (12), as shown in Fig. 7-1. A coil
can be made by winding a fine stainless steel or Nichrome wire
around a small hypodermic needle. The coil should fit snugly inside
the tube so that it will not slide when the tube is centrifuged. The
coil has many points of contact with the glass wall. Capillary wetting
action at these points of contact helps keep the trapped fractions in
the tube.

Traps can be packed with impregnated solid support material (13)
or with charcoal (14). Good trapping yields are reported, and evaporation in the mass spectrometer is controlled (see Chapter 5). Because

FIG. 7-1   Wire coil in melting point tube.

the sample must be removed from the particles for most other spectrometric analyses, however, this method is not generally applicable.

Figure 7-2 shows a device for multiple collections. Most of the
sample condenses in the Teflon tube, A. The tube and collector are
dropped into a small test tube, which is then corked to prevent sample
evaporation during centrifugation. After material is gathered in the
capillary area, C, it can be easily drawn into a microsyringe for
transfer. For sample sizes of 5-25 mg, the sample collector can be
made entirely from 5-mm tubing. A small indentation is made about
halfway up the collector so that the Teflon tubing does not slide to the
bottom during centrifugation. The sample collectors can be used with
0.03-in.-i.d. open tubular and 0.1- and 0.2-in.-i.d. packed columns.
Yields of 70-80% are obtained with oxygenated monoterpenes and
materials of similar dew points (11).

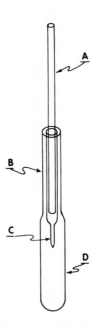

FIG. 7-2   Small collector. A, Teflon tubing. B, 5-mm-o.d.
           Pyrex tubing. C, 1-mm-i.d. Capillary. D, 7 mm o.d.

Figure 7-3 shows a trap for 25- to 100-mg samples. Stainless steel wool, B, is heated rapidly by the hot effluent gases. A temperature gradient is quickly established to keep aerosol formation to a minimum. Condensed material drips down into the glass collector, E. At the end of collections, the entire trap assembly is centrifuged to collect the trapped material in the narrow diameter region, E. The small amount that adheres to the stainless steel wool can be rinsed with carbon tetrachloride for IR or NMR examination. If a standard taper joint, D, is used, this collector can be connected directly to a vacuum transfer system (see Fig. 7-12). Vacuum transfer permits use of this trap with smaller amounts.

Figure 7-4 shows a trap for samples up to 10 g (7). This type of trap can be cooled for low boiling compounds. A drying tube can be easily attached to the outlet joint, I, to reduce collection of moisture.

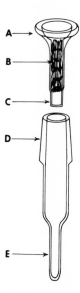

FIG. 7-3   Medium collector. A, 12/5 Socket joint. B, Fine
stainless steel wool. C, Slight restriction.
D, 14/20 Inner joint. E, 5-mm-o.d. Pyrex tubing.

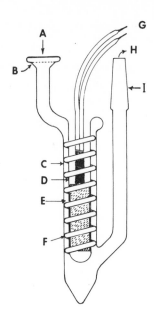

FIG. 7-4   Large collector. A, Gas in. B, 12/5 Socket joint.
C, Cold wall. D, Hot wall. E, Annular space.
F, Glass rod helix. G, Cartridge heater. H, Gas
out. I, 10/30 Inner joint.

Small amounts of low boiling compounds can be vacuum-transferred
to smaller containers or directly to spectrometer cells. Such procedures should be used only for special occasions because vacuum
transfers are tedious and time-consuming. Excessive use of stopcock
grease must be scrupulously avoided as appreciable quantities of
sample will dissolve in exposed grease.

In special cases, total eluate traps can be used for complete trapping. With one type, carrier gas is condensed with the sample
(15,16); with another, the entire effluent is captured in an evacuated
flask (17). Argon or carbon dioxide can be used as condensable carrier gas. Carbon dioxide is preferred because it is easily condensed
with liquid nitrogen and its evaporation is more easily controlled.

After the carrier is volatilized, the sample is ready for MS. For
IR and NMR analyses, the trap is rinsed with some carbon tetrachloride, and the solution is then transferred to an appropriate cell with a
microsyringe. Nearly quantitative recoveries are claimed for this
method in the milligram range. Capturing the total effluent in an
evacuated flask is a convenient method for MS analysis because the
trap can be removed from the chromatograph and then connected
directly to a mass spectrometer inlet system. For IR, NMR, and
Raman analyses, the material must be condensed and transferred.

B. Sample Manipulation

Capillary wetting action is beneficial in keeping liquid in trapping
tubes, but it is detrimental in transferring samples to cells. Figure
7-5 shows small amounts of liquid in capillary tubing. The upper
diagram shows a plug in the middle. As the hypodermic needle is inserted, the liquid will creep up outside the needle by capillary action.
Very little material will remain in the middle to be withdrawn. However, if the tube is tilted so that the liquid is at the end of the tube
(middle diagram) then very little is lost. For IR and NMR analyses,
several small portions of carbon tetrachloride can be used for
quantitative transfer to spectrometer cells.

If the trapped material in a glass tube is to be stored, the tube
should be sealed at both ends with a torch. Care should be taken that
the flame is never near the open ends. Enough water from a flame
can be condensed in a capillary to interfere in MS analyses. If neat
liquid is needed, the collected material can be centrifuged to a sealed
end. For this type of manipulation, the excess glass should not be
pulled off. After material is collected at one end, the tube is cut just
above the liquid (lower diagram, Fig. 7-5). If excess glass is not
pulled off during sealing, this length of tubing conveniently serves as
a handle during transfer of samples to hypodermic needle syringes.

Various devices (18,19) have been designed for collection of components separated by thin layer chromatography (TLC). Compounds

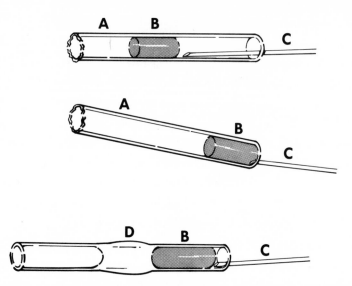

FIG. 7-5   Transfer from capillary tubes. A, Pyrex melting point
            tube. B, Organic liquid. C, Hypodermic needle. D, Seal.

are removed from adsorbents by washing with minimum volumes of
solvents. If compounds have sufficiently high boiling points, then
solvents can be removed with rotary evaporators. If they have low
boiling points, then a short preparative GC column can be used for
obtaining solvent-free material. If all adsorbent particles are not
removed for NMR and Raman studies, serious deterioration of spectra
will result. If separated materials are very polar, they will be tightly
adsorbed. However, if the adsorbent is deactivated with water, most
compounds can be quantitatively extracted.

III. NUCLEAR MAGNETIC RESONANCE (NMR) SPECTROMETRY

Since NMR spectrometry is the least sensitive of spectral methods,
special effort must be made to reduce sample size requirements.
Techniques and applications useful in examination of small samples by
high resolution NMR are discussed by Lundin et al. (20). Use of
microcells and computers has made NMR useful in flavor research.

A. Microcells

To obtain maximum signal, all of the sample must be in the detection zone. In practice, a volume of 200 µl in a standard cell (5-mm-
o.d. tube) yields good resolution, but a considerable portion of the
sample is not in the zone of the receiver coil of the probe. Several
microcells are commercially available. Some have solvent limitations;
others are difficult to fill and clean. However, the main objection to
most microcells is that high resolution spectra with 100-mHz spectrometers cannot be obtained. Frei and Niklaus (21) have proposed a
microcell consisting of a small bulb held by a Teflon chuck in a standard tube. This microcell yields excellent spectra (22).

The modified Frei-Niklaus (MFN) microcell, shown in Fig. 7-6, is
commercially available (23). Each sample bulb, G, must be checked
for resolution with tetramethylsilane (TMS). With selected bulbs,
MFN microcells will yield spectra with resolutions equivalent to those
obtained with standard cells.

The position of the bulb must be carefully checked for optimum
coupling of the sample to the receiver coil. The carbon tetrachloride
solution must be well up the stem (F, Fig. 7-6). The bulb must be at
least 0.25-in. away from the Teflon chuck, E. If the bulb is too close
to the chuck, the sensing coil is affected, homogeneity of the magnetic
field is spoiled, and resolution deteriorates. Because the entire
sample is held in the sensing zone, no particles can spin out of the
sensing zone as happens in standard cells. Hence, foreign particles
seriously damage resolution and must be removed.

B. Sample Manipulation

Every sample should be filtered. All equipment used for transferring and filtering should be thoroughly rinsed with clean carbon
tetrachloride. The area in which microcell work is done should be as
clean as possible, preferably dustless. Samples can be transferred
from trapping devices with several portions of carbon tetrachloride,

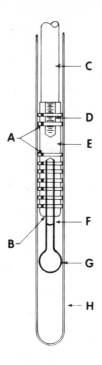

FIG. 7-6   Modified Frei-Niklaus micro-NMR sample cell. A,
            Longitudinal groove. B, 1.5-mm Hole. C, Kel-F
            rod. D, Thread. E, Teflon holder. F, 1.6-mm-
            o.d. Tubing. G, 4-mm-o.d. Sphere. H, Standard
            NMR sample tube.

about 40 μl total. The required amount of TMS is added, and the solution is drawn into a 50-μl syringe fitted with a Flath-Lundin filter set
(24). The solution is forced through fine filter paper into a microsam-
ple bulb. Capillary action in the stem may prevent filling of the bulb
directly. Liquid in the stem can be shaken into the bulb by a whipping
motion of the hand and arm. The bulb can be filled by repeatedly
filling the stem and forcing the liquid down. All air bubbles must be
eliminated before the bulb is placed in the chuck.

The chuck and bulb are then positioned in a standard NMR tube.
Excessive vertical motion of the chuck in the standard cell will cause

small flakes of Teflon to fall and wedge between the microcell bulb and
the wall of the standard tube. Such particles can seriously damage
resolution. Therefore, the chuck should be brought down in one motion
to a determined position. Then the positioning rod (C, Fig. 7-6) is
removed. A jig for positioning the spinner with respect to the sample
bulb is useful. The position of the spinner is critical because it determines whether the sample bulb is in the sensing zone or not.

With selected bulbs and properly tuned 100-MHz instruments, it is
possible to obtain usable spectra with less than milligram quantities
(20). Figure 7-7 shows a 100-MHz single-scan spectrum obtained
with 0.45 mg of humulene (22). Even the peaks from the allylic proton
(a), a sextuplet centered at 5.5 ppm, have sufficient signal-to-noise
ratio to be clearly discernible. Figure 7-8 shows 100-MHz single-scan
spectra of nootkatone (25). Since the vinyl proton signals, those from
the single proton next to the carbonyl and from the two protons on the
terminal methylene, are not split, a usable spectrum can be obtained
with less material than with humulene. The lower spectrum was
obtained with 3 mg of nootkatone in a 200-μl solution in a standard cell.
The upper spectrum was obtained with 200 μg in a 30-μl solution in a
microcell.

C. Computer Techniques

Figure 7-9 shows two spectra of isopulegol (26). In spectrum B,
the carbinol proton peaks at 3.3 ppm are not distinguishable from noise.
Spectrum A is the result of multiple scans added with a computer (26).
The signal-to-noise ratio increases as the square root of the number of
scans added (20). After 210 scans, the carbinol proton peaks are
clearly discernible.

Figure 7-10 shows spectra obtained with a 9-μ g sample of humulene.
Spectrum A, the result of 600 scans, shows peaks from solvent impurities. To eliminate the extraneous peaks, the solvent alone was also
scanned 600 times. The solvent spectrum was subtracted from the
solution spectrum with a computer to yield spectrum B. Because of

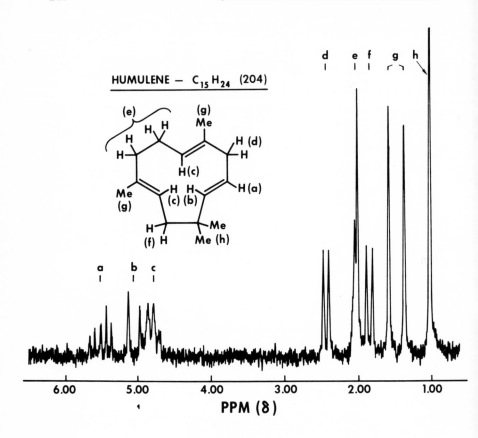

FIG. 7-7   NMR of humulene; 0.45-mg Sample in MFN microcell.

base line drift, several more scans were taken, and spectrum C was
obtained. The quality of spectrum C is not equal to that of the spectrum in Fig. 7-7, but it is usable for identification purposes. The
inset, D, shows 1700 scans over the olefinic region.

Table 7-1 summarizes minimum sample size requirements for
500-Hz nuclear magnetic resonance scans (20). Use of microcells and
multiple scanning with computers can result in usable spectra with
tens of micrograms. The Fourier transform technique (20) can lower
the sample size requirement even more. At present, lack of adequate

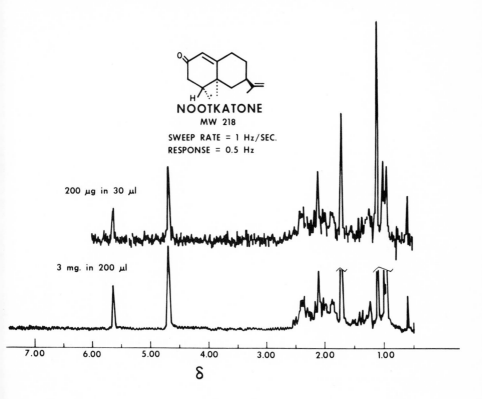

FIG. 7-8   NMR of nootkatone; 3 mg in 200-µl solution in standard cell, 200 µg in 30-µl solution in MFN microcell.

sample manipulation techniques prevents routine use of NMR with
microgram samples.

NMR spectrometry has not been widely used in flavor research for
two primary reasons: 1) Sample size requirements have been too
large for most flavor research problems, and 2) most attention has
been given to relatively simple compounds. However, as focus is
turned to more complex molecules, more spectral data are needed.
Since instrumentation and techniques have been improved, NMR can
be used in flavor research, and investigators must be familiar with

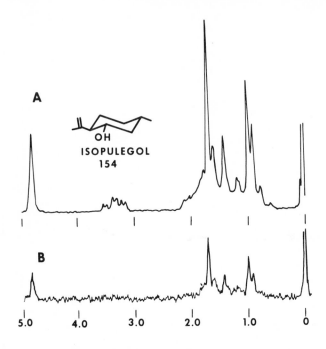

FIG. 7-9   NMR of isopulegol, HA-60. A, Time-averaged for
            210 scans. B, 0.8 mg in 50 μl, single scan.

this powerful method. Some general references are listed
(27-30).

IV. INFRARED (IR) SPECTROMETRY

IR, the most widely used spectrometric method, is a powerful tool
for determination of functional groups such as hydroxyl, amino,
carbonyl, double bonds, etc. Most investigators have access to bench-
type spectrometers. The accuracy and sensitivity of such instruments
are adequate for most flavor research problems. For best resolution
and lowest limit of detection, expensive research-type instruments
must be used (31).

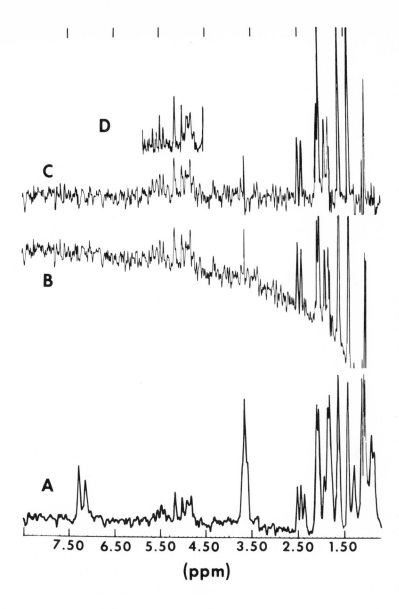

FIG. 7-10 NMR of humulene; 9 μg in 30 μl in MFN microcell.
A, Time-averaged for 600 scans. B, Same as A except solvent impurities spectrum removed. C, Same
as B except phasing corrected. D, 1700 Scans over
olefinic region.

TABLE 7-1

Minimum Sample Required for 500-Hz NMR Scans
(In Micrograms for molecular weight ca. 200)

Instrument	A-60	HR-60	HA-100
Standard sample tube (200-250 μl)	4×10^3	1.3×10^3	0.75×10^3
MFN spherical microcell or 1-mm capillary with 4-μl sample	1.5×10^3	0.5×10^3	0.3×10^3
Plus time averaging			
15 hr (one night)	65	20	15
24 hr	50	17	10
Plus Fourier transform technique	5	2	1
or plus 25 point computer least-squares smoothing	10-20 (?)	3-6 (?)	2-4 (?)

A. Gas Chromatography-Infrared (GC-IR)

Some attempts have been made to combine gas chromatography and infrared spectrometry. Standard commercial instruments have been modified (31,32) to utilize 10-cm path, 1-cc volume, flow-through gas cells. With packed columns (0.1-in.-i.d., 20-cc/min flow rate) the cell holds a 3-sec portion of a peak. Standard instruments scan in 5-10 min. Thus, after a sample is trapped in a gas cell, all subsequent peaks must be bypassed until the trapped sample is flushed out of the cell. Scott et al. (33) introduced the interrupted elution technique. Carrier gas flow is stopped while an IR spectrum is recorded. The flow is reestablished to elute the next band. This process is repeated automatically. In spite of these attempts, the use of standard, slow-scan IR instruments for GC-IR is not widely accepted.

A relatively inexpensive instrument (34) utilizes an optical filter wheel to scan from 4000 to 690 cm^{-1} in 5-12.5 sec. Three gas cells of 3-, 6-, and 9-cc volumes are available. The 6-cc cell has an optical path of 20 cm. Because this is a single-beam, purged

instrument, the base line is uneven. The short scan time requires a
fast recorder. Spectra are presented in three segments with overlap-
ping wave number intervals. Resolution is low, but it is tolerable for
qualitative information. The main objection to this instrument is that
its sample size requirement is about 0.5 mg per peak. This amount
is about two orders of magnitude too large for most aroma research
problems.

   The most promising approach in GC-IR is based on the Michelson
interferometer (35). A computer is needed to produce conventional
spectra via Fourier transformations. Because narrow slits are not
necessary for high resolution, radiation throughput is large. Therefore, sensitivity is much better than with conventional IR instruments.
The interferometer-type instrument can be adjusted for high resolution or for high sensitivity. Several minutes are needed for high
resolution spectra. Low resolution spectra can be obtained in 1-5 sec
with microgram samples. The low resolution spectra are claimed to
be comparable to those obtained with bench-type instruments. The
main disadvantage of this type of instrument is that a computer is
required for the output of conventional spectra. This type of IR instrument with the required computer costs about $75,000. Low and
Freeman (36) have reported preliminary GC-IR work with this type of
instrument. It appears that IR spectra can now be obtained as fast as
MS with similar sample sizes. The combination eliminates sample
manipulation problems. Because of the information yielded, flavor
research investigators need this GC-IR combination.

B. Sample Manipulation

   Until GC-IR combinations become more accessible, investigators
must rely on conventional instruments with batch sampling.
Fortunately, with microcells and beam condensers, usable spectra
are obtainable with 10- 50-μg samples.

   Standard 0.1-mm path cells require 50-55 μl of solution. "Microcells" from IR instrument manufacturers require 10-20 μl. These

cells have such large dead volumes that sample requirements are large and solution recovery is difficult.

Several GC-IR microcell systems for batch samples are commercially available. One is simply a small cavity cell (37) of 0.2- to 0.4-μl volume. GC-purified material can be centrifuged into it. A beam condenser is required. Spectra of neat samples can be obtained with amounts normally considered to be microsyringe needle holdup. Another microcell has an amalgamated spacer (38). All of the 0.1 X 1.0 X 10-mm cell volume is in the radiation beam. Neat or solution samples can be centrifuged into this cell. The straight-through design makes this microcell easy to clean.

Sample manipulation can be simplified by adding liquids or solids as carriers. Carbon tetrachloride vapor can be mixed with GC effluents (38). Trapped carbon tetrachloride solutions can be centrifuged into microcells. GC effluents may be condensed onto powdered KBr, which is pressed into a disk for analysis. The KBr technique is recommended only for compounds with boiling points of 150° C or higher (38). With a liquid or solid carrier technique, usable spectra can be obtained with 50-μg samples.

Sometimes solvents not suitable for IR analyses are used for extraction or transfer. Such solvents can be evaporated from the tip of a KBr cone. The sample is concentrated at the tip. The tip is cut off and is compressed into a disk (39).

Curry et al. (40) have developed a method for concentrating high boiling compounds. Trapped material is dissolved, and the solution is taken into a microsyringe. Enough solution is forced out of the needle to form a little drop on the end. When this wet needle is gently dipped into KBr powder, a small ball of KBr is collected. As solvent evaporates, more solution is forced down the needle. When the solvent is completely evaporated, the small ball of KBr, now containing the sample, is transferred to a pellet press, and a KBr disk is prepared. Usable IR spectra are reported with 1- to 10-μg samples.

There are several disadvantages in using KBr. Preparation of many samples would require prohibitive amounts of time, and samples cannot be easily reclaimed for subsequent analyses. The ionic environment may cause band shifting and broadening.

Price et al. (41) have described a microcell which is basically a capillary drilled in sodium chloride. Equipment for sealing and precise alignment is described. Usable spectra are obtained with several micrograms in carbon tetrachloride solution.

With commercially available equipment and with standard techniques, spectra are routinely available with 50- to 100-μg samples. With special equipment and techniques, spectra can be obtained with microgram quantities. With very small samples, routine, frequent analyses are not possible because of experimental time required. GC-IR systems based on the Michelson interferometer look promising for flavor research problems.

V. RAMAN SPECTROMETRY

The Raman effect has been known since 1928 (42), but it has remained in the realm of spectrometrists because of sample size limitations and other experimental difficulties. The advent of lasers and other improvements have reduced sample size requirements from grams to less than milligrams (43, 44).

Raman spectrometry yields information complementary to IR; i.e., bands forbidden in IR appear in Raman spectra and vice versa. Thus, Raman is of value in structure determinations of functional groups which show only weak absorptions in IR, such as tetrasubstituted double bonds, triple bonds, and carbon-sulfur and sulfur-sulfur bonds. Because of insufficient use, Raman spectrometry is still only a potential tool. In time, however, it should become very useful and take its place as one of the major spectrometric methods.

Figure 7-11 shows a Raman spectrum of β-bisabolene (45). The band in the 760-cm^{-1} region is due to some form of ring mode

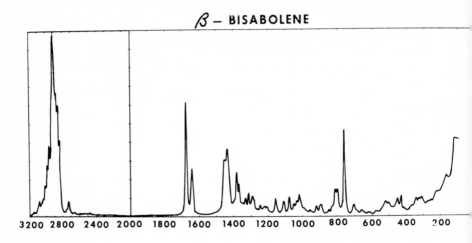

FIG. 7-11   Raman spectrum of β-bisabolene.

vibration associated with substituted cyclohexene rings. The band in the 1643-cm^{-1} region is due to the C=C stretch of the disubstituted double bond. The band in the 1675-cm^{-1} region is due to the C=C stretch of the trisubstituted double bonds. The ratio of intensities indicate two trisubstituted double bonds versus one terminal methylene double bond.

Sensitivities of commercial Raman instruments approach those of IR instruments. Usable spectra are obtained with quantities of 0.05-0.1 μl or less (43, 44).

Figure 7-12 shows a simple vacuum transfer device useful for filling micro-Raman sample tubes (45). Glass capillaries, E (0.1- to 0.4-mm i.d., 20 mm long), sealed at one end, are placed open end down in the constricted tip of container D. Most of the capillary sample tube must stand away from the wall of the container, D. If most of the capillary tube is touching the wall, most of the sample will adhere to the outside wall by wetting action. High vacuum stopcock grease must be carefully applied only to the lower areas of standard taper

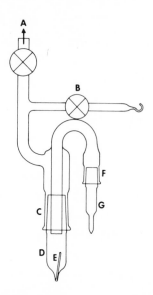

FIG. 7-12   Vacuum transfer system. A, To high vacuum mani-
             fold. B, Stopcock to release vacuum. C, 24/40
             Joint. D, Microcell holder. E, Microcell. F,
             14/20 Joint. G, Sample container.

joints, C and F. If excess grease is exposed at the upper edges of
either joint, considerable amounts of sample will dissolve in the
grease.

After sample and microcell containers are connected to the transfer system, both stopcocks, A and B, are closed. The sample, in G,
is thoroughly cooled with liquid nitrogen. Then stopcock A is opened
to connect the transfer system to a high vacuum manifold. When the
system has reached 10^{-5} Torr, stopcock A is closed and the sample is
gently warmed. After dissolved air is released, the sample is cooled
again with liquid nitrogen. The released air is pumped away by the
high vacuum system. This sequence is repeated several times to remove most of the air in the sample. Then the microcell container, D,
is cooled with liquid nitrogen, and the sample in G is warmed. The

entire sample is collected in the small tip of D by gently warming the
larger, upper area and by keeping the tip cool. The sample in the tip
is permitted to warm up just enough to liquefy. At this moment,
stopcock B is opened, and atmospheric pressure forces liquid into the
microcell. This vacuum transfer system permits quantitative transfer
of milligram quantities of liquids, free of any particulate matter
which may have contaminated the sample·in container G. Fortunately,
most compounds encountered in flavor research are liquids. Solids
are usually handled in solution. Some needle-like crystals have been
inserted into the capillary tubes, and usable spectra have been
obtained. Diglyme is a convenient solvent since it exhibits only a few
Raman bands of low intensity and since it has a sufficiently high
boiling point to be used with the vacuum filling technique. Manipulation of solids into the Raman microcells is not as easy as the liquid
vacuum transfer method.

VI. SUMMARY

Improvements in instruments and techniques permit acquisition of
usable spectra from microgram quantities. However, so much time
and effort are required for analyses of very small samples that large
numbers of spectra cannot be obtained in a reasonable time. Sample
manipulation can be eliminated with GC-IR, but combinations of GC
with NMR and Raman seem unlikely as yet. Fortunately, a great
number of identifications can be accomplished with GC-MS (see Chapter 6). Tedious and time-consuming sample manipulations can be
performed on those few unidentified fractions which appear to be
sensorially important. Further developments in application of batch
spectrometric analyses depend on the ingenuity of investigators in
developing techniques for manipulation of microgram quantities
without loss or contamination.

REFERENCES

1. A. Wehrli and E. Kovats, J. Chromatog., 3, 313 (1960).

2. E. P. Atkinson and G. A. P. Tuey, in Gas Chromatography 1958 (D. H. Desty, ed.), Butterworths, London, 1958, p. 270.

3. P. Kratz, M. Jacobs, and B. M. Mitzner, Analyst, 84, 671 (1959).

4. A. E. Thompson, J. Chromatog., 6, 454 (1961).

5. W. D. Ross, J. F. Moon, and R. L. Evers, J. Gas Chromatog., 2, 340 (1964).

6. H. Schlenk and D. M. Sand, Anal. Chem., 34, 1676 (1962).

7. R. Teranishi, J. W. Corse, J. C. Day, and W. G. Jennings, J. Chromatog., 9, 244 (1962).

8. S. Dal Nogare and R. S. Juvet, Jr., Gas-Liquid Chromatography, Wiley (Interscience), New York, 1962, p. 295.

9. K. Biemann, Mass Spectrometry, Organic Chemical Applications, McGraw-Hill, New York, 1962, p. 32.

10. R. L. Hoffman and A. Silveira, Jr., J. Gas Chromatog., 2, 107 (1964).

11. R. Teranishi, R.A. Flath, T. R. Mon, and K. L. Stevens, J. Gas Chromatog., 3, 206 (1965).

12. R. G. Buttery, USDA, Albany, Calif., private communication, 1969.

13. J. W. Amy, E. M. Chait, W. E. Baitinger, and F. W. McLafferty, Anal. Chem., 37, 1265 (1965).

14. J. N. Damico, N. P. Wong, and J. A. Sphon, Anal. Chem., 39, 1045 (1967).

15. P. A. T. Swoboda, Nature, 199, 31 (1963).

16. I. Hornstein and P. Crowe, Anal. Chem., 37, 170 (1965).

17.   S. Dal Nogare and R. S. Juvet, Jr. , Gas-Liquid Chromatog. ,
      Wiley (Interscience), New York, 1962, p. 254.

18.   J. H. Nagel and J. C. Dittmer, J. Chromatog. , 42, 121 (1969).

19.   Quickfit, Inc. , Fairfield, N. J.

20.   R. E. Lundin, R. H. Elsken, R. A. Flath, and R. Teranishi,
      Appl. Spectry. Rev. , 1(1), 131 (1967).

21.   K. Frei and P. Niklaus, private communication, 1964.

22.   R. A. Flath, N. Henderson, R. E. Lundin, and R. Teranishi,
      J. Appl. Spectry. , 21, 183 (1967).

23.   Kontes Glass Co. , Vineland, N. J.

24.   Hamilton Co. , Whittier, Calif.

25.   R. E. Lundin, USDA, Albany, Calif. , unpublished data.

26.   R. E. Lundin, R. H. Elsken, R. A. Flath, N. Henderson,
      T. R. Mon, and R. Teranishi, Anal. Chem. , 38, 291 (1966).

27.   F. A. Bovey, Nuclear Magnetic Resonance Spectroscopy,
      Academic, New York-London, 1969.

28.   L. M. Jackman and S. Sternhell, Applications of Nuclear Mag-
      netic Resonance Spectroscopy in Organic Chemistry, 2nd ed. ,
      Pergamon, New York, 1969.

29.   J. W. Emsley, J. Feeney, and L. H. Sutcliffe, High Resolution
      Nuclear Magnetic Resonance Spectroscopy, Vols. 1 and 2,
      Pergamon, New York, 1965, 1966.

30.   R. Teranishi, R. E. Lundin, W. H. McFadden, and J. R.
      Scherer, in The Practice of Gas Chromatography (L. S. Ettre
      and A. Zlatkis, eds.), Wiley (Interscience), New York,
      1967, p. 407.

31.   Wilks Scientific Corp. , South Norwalk, Conn.

32.   J. D. Propster, Pittsburgh Conference on Analytical Chemistry
      and Applied Spectroscopy, March 1965.

33.   R. P. W. Scott, I. A. Fowlis, D. Welti, and T. Wilkins, in
      Gas Chromatography, Rome 1966 (A. B. Littlewood, ed.),
      Bartholomew Press, Dorking, Surrey, England, 1966, p. 318.

34. Beckman Instruments, Inc., Fullerton, Calif.

35. M. J. D. Low, Anal. Chem., 41(6), 97A (1969).

36. M. J. D. Low and S. K. Freeman, Anal. Chem., 39, 194 (1967).

37. Barnes Engineering Co., Stamford, Conn.

38. Carle Instruments, Inc., Fullerton, Calif.

39. Wickstik, Harshaw Chemical Co., Cleveland, Ohio.

40. A. S. Curry, J. Read, C. Brown, and R. Jenkins, J. Chromatog., 38, 200 (1968).

41. G. D. Price, E. C. Sunas, and J. F. Williams, Anal. Chem., 39, 138 (1967).

42. C. V. Raman, Indian J. Phys., 2, 387 (1928).

43. G. F. Bailey, S. Kint, and J. R. Scherer, Anal. Chem., 39, 1040 (1967).

44. S. K. Freeman and D. O. Landon, Anal. Chem., 41, 398 (1969).

45. J. R. Scherer, USDA, Albany, Calif., private communication, 1969.

Chapter 8

CORRELATION WITH SENSORY PROPERTIES

I. INTRODUCTION: SOME PROBLEMS

The major challenge in modern flavor research lies in achieving correlation of sensory quality with chemical identity. Due to remarkable development of the analytical techniques that have been discussed in the previous chapters, the chemistry of volatile constituents can be determined, though in certain cases much patience, care, and skillful use of sophisticated and expensive procedures are required. There has, however, been no revolutionary deveooment of techniques of sensory evaluation; and yet the number of substances needing to be evaluated as possible contributors to any given flavor has become staggering. Moulton (1) has suggested that the complex array of substances in most odors, or even in the air about us, presents our odor receptors with a task not unlike that of trying "to determine visual thresholds during a fireworks display."

Human judgment must be used in conjunction with gas chromatographic separations and subsequent identification of components to determine which (if any) contribute to the characteristic aroma being studied. Human judgment must also describe the intensity and quality of individual peaks or groups of peaks and determine their relationship to the aroma of the whole isolate. In short, a bioassy of aroma based on the stimulation of the human nose must be carried out. It would help greatly if knowledge of the basis of olfaction and taste were available. Since it is not available, we must solve the problem of correlating chemical structure with flavor the best we can. As we said in Chapter 1, the situation is very much like looking for one particular needle or a handful of particular needles, in a haystack full of needles!

Let us assume that some very good chromatograms have been obtained of the head space vapor over a foodstuff. There are perhaps 80 or 90 peaks present. Since a blank sample has been analyzed under the same conditions and only a few minor peaks found, one is quite certain that the head space chromatogram contains a minimum number of artifacts and that it rather closely represents the natural composition (as revealed by the particular column used) of the vapor over the product — and, of course, this vapor exhibits the characteristic aroma we hope to understand.

So far so good: But it is quite possible that the head space chromatograms have not shown all the compounds that must be present together to produce the characteristic aroma. Perhaps the vapor pressure of certain indispensable constituents of the aroma is too low to provide enough molecules for gas chromatographic detection — even though a sufficient number are present for us to smell. Some also may not have been eluted from the columns.

Because of this possibility — and because a sufficient quantity of individual constituents is needed to allow their unequivocal chemical identification — an aroma concentrate from the foodstuff has been prepared using the mildest conditions possible and making every effort, through use of glass apparatus and highly purified solvents and reagents, to avoid formation of artifacts or at least to keep their formation minimal.

Chromatograms of the concentrate obtained on the same column as the head space chromatograms show several hundred peaks. A significantly greater number of relatively high boiling compounds are detected than in the head space. Fewer low boiling components are detected than in the head space and, of course, the quantitative relationships between constituents are quite different from those in the head space because of the effect of the isolation procedures used. A number of the peaks are undoubtedly artifacts. One hopes, of course, that the GC of the concentrate will be as similar as possible to that of the head space.

Again, one wonders whether all the indispensable aroma compounds are present, i.e., those necessary for production of the characteristic aroma. This time a good portion of these compounds have probably been removed from the food, but it is possible that some have not been eluted from the column and hence not detected.

At this point sensory evaluation is very badly needed as a guide for the chemical work. Reference to Table 8-1, which outlines the situation, shows that "nasal appraisals" are needed at the following places (marked with an asterisk):

(a) The headspace sample — is its aroma typical and of good quality?

(b) The aroma concentrate — if one dilutes it to its original concentration, is its aroma characteristic of the product? If it is

not, is one justified in going on and studying its chemical
composition?

This last is not a particularly easy question to answer. One view is
that under such conditions (absence of characteristic aroma in the
concentrate), without strong evidence to the contrary, time should not
usually be spent identifying its constituents, because the biological
property (the aroma) which is the primary reason for the work is no
longer present.

At appropriate dilution the aroma concentrate should resemble the
aroma of interest even though its composition varies from that of the
head space. The concentrate will certainly not smell exactly like the
head space but it should definitely suggest the product being studied.
If possible, several methods for preparation of aroma concentrates
should be investigated to learn how best to optimize aroma quality.
One must find a practical balance between amount of concentrate
isolatable and odor quality.

An example of an investigation in which this was done concerned
the volatiles from Delicious apple essence (2). As shown in Table
8-2, yields were determined for aroma concentrates isolated by five
different methods: isopentane, diethyl ether, carbon dioxide and
fluorocarbon extractions, and adsorption on charcoal. All five
isolates plainly exhibited strong apple aroma, but careful panel evalu-
ɛtions were carried out. Thresholds of original essence and of each
apple extract were determined and compared to see whether a major
portion of indispensable components had been isolated. As shown in
Table 8-3, ratios of these thresholds and ratios of weight of essence
used to weight of each isolate obtained were quite similar. This
indicated that the important aroma substances were present in every
extract. The relative quality of the extracts in comparison with the
aroma of freshly pressed apple juice was evaluated by the triangle
method and by the paired standard method. Each extract was somewhat different in aroma from the standard fresh juice, but the

TABLE 8-1

Some Considerations of Aroma Evaluation and Gas Chromatographic Separations

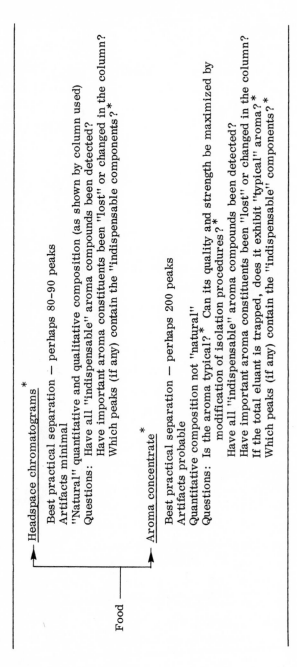

Food

Headspace chromatograms *

    Best practical separation — perhaps 80-90 peaks
    Artifacts minimal
    "Natural" quantitative and qualitative composition (as shown by column used)
    Questions: Have all "indispensable" aroma compounds been detected?
            Have important aroma constituents been "lost" or changed in the column?
            Which peaks (if any) contain the "indispensable components?*

Aroma concentrate *

    Best practical separation — perhaps 200 peaks
    Artifacts probable
    Quantitative composition not "natural"
    Questions: Is the aroma typical?* Can its quality and strength be maximized by
            modification of isolation procedures?*
            Have all "indispensable" aroma compounds been detected?
            Have important aroma constituents been "lost" or changed in the column?
            If the total eluant is trapped, does it exhibit "typical" aroma?*
            Which peaks (if any) contain the "indispensable" components?*

* "Nasal appraisal" needed.

TABLE 8-2

Extraction of Volatiles from Aqueous Apple Essence[a]

Extractant	Isopentane	Ether	Charcoal[b]	CO_2	$CClF_2-CClF_2$
Yield, g from 3 kg aqueous essence	2.53	6.05	8.41	2.96[c]	1.36
Residual solvent, %	55	40	44	< 1	~ 10
Calculated net yield, g per kg aqueous essence	0.38	1.21	1.58	0.98	0.41

[a]Reprinted from Ref. (2), p. 280, by courtesy of the Institute of Food Technologists.
[b]With elution by ether.
[c]Calculated from actual yield of 9.55 g from 9.67 kg of aqueous essence.

TABLE 8-3

Ratios of Thresholds and Weights for
Apple Aroma Concentrates[a]

	Extractant			
	Isopentane	Ether	Charcoal	Fluorocarbon
Threshold ratio essence/extract	1060	460	340	2150
Weight ratio essence/extract	1190	495	357	2210

[a] Reprinted from Ref (2), p. 282, by courtesy of the Institute of Food Technologists.

isopentane and ether extracts were most like it and were found not to differ significantly from it. They were, therefore, the most desirable isolates for further detailed examination — even though they had not been obtained in the greatest yields (Table 8-2)! The differences in yields were largely caused by the degree to which lower boiling alcohols had been extracted.

Although it is hard work to accumulate sufficient quantities of flavor concentrates to permit their sensory evaluation by a panel, it is not beyond reason to do so. However, when one considers carrying out aroma evaluations of individual peaks in chromatograms, each of which amounts to micrograms or fractions of micrograms, one's enthusiasm fades in the face of the work required. Alternative approaches are hopefully sought.

One alternative is to isolate and identify all the peaks and their various components (because many peaks will contain mixtures) in order to reproduce the composition of the aroma in question. In many ways this procedure is least attractive. It should be carried out only when sensory evaluation has not been helpful in selecting peaks that should have priority in the identification work. Life is already much too short to spend valuable time identifying many organic substances that are probably already known and not important to the biological property in question — the aroma and flavor.

A good bioassay method is needed by flavor chemists. If one were engaged in the isolation and identification of a drug or other biologically active material, one would depend heavily on bioassay until the active principle or principles were identified, and then chemical or physical assays would be developed. Somehow one tends to feel safer with bioassays that use the gain or loss of weight of test animals or the physiological state of rat livers as evidence of biological activity than one does at the prospect of asking impressionable, opinionated, and unreliable human beings to judge aroma in isolates. This attitude must be overcome and very serious efforts made to use sensory evaluation to select the gas chromatographic peaks of importance, thus avoiding the necessity of identifying all components and then determining their flavor contribution.

II. DIRECT ASSESSMENT OF FLAVOR SIGNIFICANCE

A. Relative Contribution to Odor Intensity Based on Odor Thresholds

One measure of a compound's flavor significance is the intensity of its aroma. This is relatively easily measured by determining its threshold, i.e., the minimum quantity of the compound that causes an odor detectable by a specified percentage of panel members. Even though the existence of sensory thresholds is open to question (3, 4), thresholds have great practical use in operational terms. In the same manner the remarkable variability of odor and taste sensitivity within and between individuals (5, 6) is acknowledged, but at the same time, average threshold values obtained from good-sized groups of individuals have much practical value.

Thresholds must be determined under carefully controlled and specified conditions. Compounds must be highly purified by gas chromatography and tested quickly before degradation or change can occur. As shown in Table 8-4, the method used for threshold determination (sniffing beakers vs. Teflon squeeze bottles) affects the values found (7). For this reason, and because methods have not been described and compounds were of unknown and probably inadequate purity, threshold values published in the literature "before gas chromatography" have little value.

If the actual concentration of an odorant in a food product is greater than its threshold concentration, it is logical to suppose that the odorant contributes significantly to that food's flavor. However, the threshold must have been determined in a medium similar in polarity to that of the food matrix. The use of water vs. oil as solvent media for fatty acid threshold determinations (8) is summarized in Table 8-5. It is clear that solubility has a major effect on results. Another

TABLE 8-4

Comparison of Olfactory Thresholds Obtained by Different Methods[a]

Compound	Thresholds, ppb in water	
	Sniffing beakers	Squeeze bottles
2-Methylpropanal	5 ± 3	0.9 ± 0.2
n-Heptanal	16 ± 6	3 ± 1
n-Undecanal	12.5 ± 5	5 ± 2
n-Nonanal	98 ± 48	1 ± 0.2

[a] From Ref. (10), p. 763.

TABLE 8-5

Flavor Thresholds for Volatile Fatty Acids in
Water and Oil[a]

Fatty acid	Water, ppm	Oil, ppm
C_2	54.0	—
C_4	6.8	0.6
C_6	5.4	2.5
C_8	5.8	350.0
C_{10}	3.5	200.0
C_{12}	—	700.0

[a] Reprinted from Ref. (8), p. 679, by courtesy
of the Institute of Food Technologists.

interesting example is that of oct-1-en-3-ol, the compound responsible
for mushroom flavor in dairy products. As shown in Table 8-6, its
flavor threshold increases 10 times as it is moved from water, to skim
milk, to butterfat as test media (9).

Based on observations with mixtures of aldehydes, a sulfide,
an amine, and an acid — each at a level below its own threshold —
Guadagni et al. (10) concluded that certain compounds have an additive
effect at subthreshold levels. This meant that even constituents present in a flavor concentrate at lower than threshold levels might have
very significant flavor or aroma effects. Working on the assumptions
that the total odor intensity of a flavor concentrate was measured by

TABLE 8-6

Flavor Thresholds of Oct-1-en-3-ol in Various Media[a]

Medium	Flavor threshold, ppb
Water	1
Skim milk	10
Butterfat	100

[a] From Ref. (9), p. 257.

its odor threshold, and that this total intensity was the sum of the intensities (or thresholds) of each of its component fractions, the relationships shown in Table 8-7 were devised to assess the odor contributions of hop oil fractions and their individual components (11). Knowledge of the chemical identity of constituents was not necessary. Results of application of these relationships to determination of relative contributions to aroma of the hydrocarbon and oxygenated fractions of hop oil (11) and carrot oil (12) are summarized in Table 8-8. They showed that the most important odor constituents were in the oxygenated fraction of carrot oil. In hop oil, on the other hand, hydrocarbons were prime targets for investigation.

Another example of the practical usefulness of the "odor unit" concept (13) in guiding one's attention toward odor-contributing constituents is a study of apple volatiles by Guadagni et al. (14). Essences prepared by common commercial procedures from Delicious apples were separated gas chromatographically. The relative importance of the various peaks in terms of characteristic apple odor was determined by a trained panel, each member of which described the characteristic odor of individual fractions as they were eluted from the column after

TABLE 8-7

Relative Contribution of Components to the Odor Intensity of
a Total Volatile Isolate[a]

T_c = threshold conc (ppm, ppb) of isolate or of a given component

U_o = odor intensity units of isolate

F_c = conc (ppm, ppb) of given component or peak in the isolate

$$\frac{F_c}{T_c} = U_{o_p} = \text{odor intensity units of component or peak from isolate}$$

$$U_{o_{p_1}} + U_{o_{p_2}} + U_{o_{p_3}} + \ldots = U_o \text{ odor intensity units of isolate}$$

$$\frac{U_{o_{p_X}}}{U_o} \times 100 = \text{per cent contribution of peak X}$$

[a] From Ref. (11), p. 142.

TABLE 8-8

Odor Thresholds and Relative Contribution of Certain Fractions to
Total Odor Intensity of Hop and Carrot Root Oils[a]

	Threshold, ppb in water	% Composition of whole oil	% Odor Contribution
Carrot root oil	6	100	100
Hydrocarbon fraction	75	58	5
Oxygenated fraction	2	30	90
Hop oil	12	100	100
Hydrocarbon fraction	15	86	69
Oxygenated fraction	5	14	34

[a] From Ref. (11), p. 142, and Ref. (12), p. 1015.

20 μl of essence had been injected. The judges could not see the
chromatogram record and were not told the nature of the sample.
This is the "nasal appraisal" method mentioned in Chapter 4, p. 127.
The results of their evaluations are shown in Fig. 8-1.

A high percentage of "apple" or "apple-like" descriptions was assigned
to peaks 9-12, indicating that these peaks made an important contribution to apple aroma. Additional evidence that this was the case was
obtained by sensory evaluation of trapped samples of the total effluent,
of combined peaks 1-8, and of combined peaks 9-13. Triangle test
comparisons of the aroma of the total effluent versus that of peaks
9-13, and of the aroma of peaks 1-8 versus that of peaks 9-13 showed
significant preference for peaks 9-13. Triangle test comparisons of
individual peaks from 9 through 13 showed that peak 11 (the smallest
on the chromatogram in Fig. 8-1!) accounted for most of the desirable
aroma. Peaks 9 and 10 were next most important and peaks 12 and
13 (the largest components!) were least important.

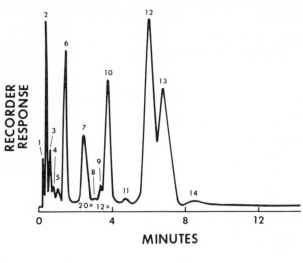

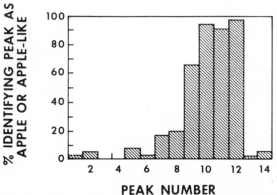

FIG. 8-1   Relative importance of various peaks in terms of
            characteristic apple odor. Reprinted from Ref. (14),
            p. 519, by courtesy of the Institute of Food
            Technologists.

The odor threshold and relative odor intensity of the apple-like
fractions were determined. They are summarized in Table 8-9 (14).
Review of these data shows that when the essence was collected in two
portions, the sum of the odor units equaled those obtained from the
total column effluent, and that almost all of the odor was in the

TABLE 8-9

Relative Odor Intensity and Thresholds of
Selected Apple Fractions[a]

Fraction	Threshold, ppm	Relative odor units, F_c/T_c[b]	
Total effluent	1	160	
Peaks 1-8	23	6 ⎫	
Peaks 9-13	0.140	154 ⎭	160
Peaks 9-13	0.170	185	
Peaks 9-10	0.390	20 ⎫	
Peak 11	0.001	160 ⎬	192
Peaks 12-13	1.120	12 ⎭	

[a] Reprinted from Ref. (14), p. 520, by courtesy of the
Institute of Food Technologists.

[b] F_c, ppm of essence equivalent in each fraction based
on a total essence concentration of 160 ppm.
T_c, threshold concentration of each fraction in ppm.
By definition, T_c equals one odor unit.

portion made up of peaks 9-13. The total odor units for individual
peaks within the 9-13 fraction was 104% of the total found experi-
mentally for the whole. Peak 11 exhibited the most intense odor
since its threshold was by far the lowest. It is noteworthy that the
smallest peak on the chromatogram had the greatest odor intensity.

Examination of chromatograms from numerous samples of apple
essences that had been held under various storage conditions resulted

in the ability to correlate their sensory quality with the total area under peaks 9-12. A useful quality control procedure was thus developed without the necessity of accomplishing chemical identification of the indispensable components.

Subsequent investigations (15), which employed coupled gas chromatography-mass spectrometry as well as infrared and proton magnetic resonance spectrometry, resulted in the identification of many apple constituents. Ethyl 2-methyl butyrate, trans-2-hexenal, and n-hexanal were identified as indispensable components of the aroma of Red Delicious apples. The amounts in which they were recovered and their olfactory thresholds are shown in Table 8-10.

Additional examples (16-18) of the correlation of sensory properties with chemical structure might have been cited. However, the above example was chosen for discussion because of the completeness with which it has been described (2, 14, 15). Knowledge of the identity of the most important components has even been combined with further odor threshold and gas chromatographic analysis to obtain material balances of these constituents during use of the WURVAC process to make apple aroma concentrates which are suitable for dehydrated food products (19).

All of the above discussion about the relative contribution made to the total odor of a mixture by certain fractions or certain individual substances is related only to intensity. Nothing has been said about odor quality, i.e., what the mixture or any one of its components smells like. Furthermore, nothing was implied about the relationship between stimulus concentration and intensity of sensation above threshold. In all of the cases examined by Guadagni and co-workers (11,12,14,15), additivity of subthreshold concentrations was observed. Other effects have been noted, however. A synergistic interaction between methyl ketones in milk solution was observed when a detectable flavor was produced even though each compound was present in an amount well below its threshold (20). In another instance,

TABLE 8-10

Calculated Recovery of Some Apple Essence Constituents[a]

Component	mg from 1 kg of 150-fold aqueous essence		Olfactory threshold, ppm v/v
	Isopentane extract	Ether extract	
Ethanol	~ 2	49	100.
Ethyl acetate	7	56	5.0
n-Hexanal[b]	40	66	0.005
Propyl propionate[b]	—	—	0.057
n-Butyl acetate[b]	39	61	0.066
Ethyl butyrate	9	13	0.001
Ethyl 2-methyl butyrate[c]	—	—	0.0001
trans-2-Hexenal[d]	145	295	0.017
2-Methylbutyl acetate[d]	30	45	0.005
n-Hexanol[d]	42	195	0.5

[a] From Ref. (15), p. 34.

[b] In peak 10.

[c] In peak 11.

[d] In peak 12.

Meijboom (21) concluded that in certain mixtures of aldehydes in paraffin oil masking effects occurred. For example, the odor of cis-3-hexenol was not detected in the presence of large amounts of

decadienal even though the hexenol was present in amounts greater
than 100 times its threshold concentration. A recent discussion of
problems of aroma interactions is presented by Keith and Powers (22).
Obviously a great deal must be learned before questions about additivity, masking, or synergism of odor intensities can be answered. In
the meantime, the "odor unit" concept is a useful tool because it
reduces all compounds to a "sensory common denominator" — that of
the odor threshold — rather than to the usual basis of concentration
alone.

B. Odor Quality

Just as important as threshold determination in achieving valid
"nasal appraisal" of flavor isolates is the ability to describe and
recognize an isolate's "characteristic" aroma quality. Flavor investigations are greatly simplified if certain volatiles in the product of
interest obviously exhibit the "characteristic aroma" of that product.
In such cases "sniffing" the effluent of a gas chromatographic column
or "sniffing" fractions from silica gel columns, etc., can provide
easy guidance to the components of greatest olfactory importance.
These can then be given primary attention and their identity can be
determined. Some examples of such "character impact compounds"
are given in Table 8-11. The odor of each substance by itself
immediately brings to mind its source product. Compounds whose
aromas suggest greenness, fruitiness, bouquet, fullness, and the like
are termed "contributory flavor compounds" (23).

A very serious complication in efforts to describe odors is the
fact that some compounds exhibit different odor qualities at different
concentrations. They may have one "character" at threshold levels
and quite another at suprathreshold concentrations. Because of this,
it is important to learn whether or how odor characteristics change
with increasing concentrations of a given pure compound. In addition,
if they are to be valid, odor comparisons must be carried out between

TABLE 8-11

Some "Character Impact Compounds"

Compound	Odor quality	Threshold, ppb in water
Ethyl trans-2-cis-4- decadienoate	Bartlett pear	—
Nona-2-trans-6-cis-dienal (25)	Cucumber	0.1
Ethyl 2-methylbutyrate (15)	Apple	0.1
2-Methoxy-3-isobutylpyrazine (26)	Green bell pepper	0.002
Nootkatone (27)	Grapefruit	170

[a] Numbers in parentheses indicate references to the literature.

comparable concentrations and under as reproducible and controlled conditions as possible.

In very many, if not most, flavor investigations "character impact compounds" are not to be found. Rather, these foods' characteristic aromas seem due to man's integrated response to a number of "contributory flavor compounds." This was the case, for example, for hop oil (11), apricots (28), and irradiated beef (18). In such a situation, sniffing the effluent of a GC column can lead to a morass of differing descriptive terms. Different people use different terms to describe the same odor, and since virtually every organic compound has an odor of some kind, it is a big problem to unscramble the resulting terminology and reach agreement. The intensive efforts by Harper et al. (29) to develop a standardized nomenclature for odors like the Munsell scheme for describing colors are discussed in their

book, <u>Odour Description and Odour Classification</u>. They employ factor
analysis and multidimensional similarity analysis. The greatest
single identifiable source of disagreement in odor descriptions
appeared to stem from the various observers' views of pleasantness or
unpleasantness of the stimuli. It will be some time before such
investigations can provide practical assistance to chemists who seek
to single out indispensable aroma constituents while essentially
ignoring the others. Recent work by von Sydow et al. (75) is of interest.

Empirical approaches to selection and identification of "contributory
flavor compounds" can quite often be successful. If direct "nasal
appraisal" of a GC column effluent is not helpful, collection of portions
of the effluent in cold traps and evaluation of their aromas can be
profitable. With luck, one can sometimes find the portion of a chromatogram that contains the indispensable odor components. In such
efforts it is well to use a relatively "poor" column as well as a good
"high resolution" column. The very good column can separate components too far apart for their "odor relatedness" to be noted, but the
"poor" column may provide guidance about which of the well-separated
peaks contain the compounds of interest. The recovery of components
by collection of either total or partial GC column effluents is rarely
quantitative. In addition, the extremely small quantity of trapped
material forces odor evaluations to be rapid and "informal" in nature.
Nevertheless, important clues about sensory contributions can be
obtained. In evaluations of this kind it is wise to evaluate "total"
effluent minus the portion of possible interest, as well as the interesting portion itself. Peaks whose absence makes a difference to total
effluent aroma are important ones. Merely to show similarity between
a peak or a group of peaks and the whole effluent is not sufficient since
extraneous influences may have caused no difference to be noted.

Selection of an appropriate medium in which to evaluate compounds
suspected to be aroma contributors is vital to success. Use of a
bland, nonodorous version of the product under study is best. In this

way compounds can be added to a "natural" food matrix in known
amounts and their thresholds and effects on odor quality can be noted.
By this means important flavor constituents of beer (30) and of irradiated beef off-odor (18) have been established. In the latter case
major peaks in an "odor-contributing" portion of GC effluent were
isolated and identified. Synthetic mixtures of these pure known compounds, based on the relative quantities indicated on the chromatogram,
were prepared in aqueous solutions and added in known amounts to
slurries of nonirradiated beef. These were compared with irradiated
slurried beef. It was obvious almost immediately that some of the
compounds contributed little or nothing to irradiation odor, while
others affected it greatly. Trial and error mixing and concurrent odor
evaluation led to a "mix" that panels could not differentiate from
irradiated slurries. The compounds of major importance, the amounts
required for off-odor production in nonirradiated slurries, and their
individual odor thresholds in the slurries and in water are shown in
Table 8-12. We have already noted the significance of subthreshold
concentrations as important contributors to aroma. Much more must
be learned before the relationship among chemical structure, concentration, and sensory effects will be understood. Selection of appropriate media in which to test these factors is necessary.

C. Olfactory Purity

All investigations of relationships between molecular structure
and olfactory quality are fundamentally dependent on an assurance
that the compound of interest has "olfactory purity," that is, that any
impurities which may be present nave no effect on the odor of the compound. In view of the extremely low thresholds of many compounds
[0.002 ppb for the bell pepper compound, for example (26)], they can
be expected in mixtures to have very significant effects at extremely
low levels. It has been pointed out in Chapter 3 that the presence of
only a fraction of a per cent of β-ionone (threshold 0.007 ppb in water)
mixed with n-amyl acetate (threshold 5 ppb in water) would have an

TABLE 8-12

Odor Thresholds of Beef Irradiation Flavor Components in Water
and in Beef Slurries and Amounts Required for Off-Flavor
Production[a]

Compounds	Individual Thresholds		Quantity for irradiation odor, ppm
	ppm in water	ppm in slurry	
Methional	5.0	6.1	3.0
n-Nonanal	0.50	0.94	0.15
Phenylacetaldehyde	0.25	> 7.6	0.30

[a] From Ref. (18), pp. 21, 24.

important sensory effect on the aroma quality of the acetate. The
need for achieving olfactory purity of compounds whose threshold and
odor quality are to be determined cannot be overemphasized. Use of
threshold determinations to measure attainment of "constant threshold
values" after repeated purification steps can be a powerful tool for
achieving olfactory purity.

III. INDIRECT ASSESSMENT OF FLAVOR SIGNIFICANCE

The evaluation of sensory importance of food constituents by any
means other than direct isolation and identification of compounds combined with direct determination of their olfactory or taste properties
is, for purposes of this discussion, defined as indirect assessment.
Therefore, the broad range of correlations of data from "objective"
analyses of many kinds with data from "subjective" olfactory analyses
falls in this category. Though the objective data can be derived from
almost any kind of instrumental analysis, they are most often

obtained in the form of numerous gas chromatographic patterns. Thus, when one speaks today of correlations between sensory and instrumental data, gas-liquid chromatography (GLC) data are the objective data that most often come to mind. In view of available knowledge of fundamental aspects of gas chromatography in relationship to flavor investigations, gas chromatographic patterns are recognized to be highly dependent on experimental conditions and the equipment used. Hence, vigorous efforts must be made to assure that the GLC data have, in fact, some realistic relationships to aroma and flavor quality.

Although human judgments always provide the ultimate test of sensory quality, particularly in terms of pleasantness, unpleasantness, and overall acceptability, it is very desirable that objective analyses be developed which can take their place. It is much easier to calibrate and learn the accuracy, precision, sensitivity, and long-term reproducibility and reliability of instruments than to establish these same characteristics for people. Thus, correlation of objective (or instrumental) techniques with sensory techniques of flavor and odor evaluation is much to be desired. Such correlations may be highly significant even though the objectively measured constituents do not reflect sensory changes at all. The objective analysis may measure the effects of other variables which also affect the sensory measurements. This type of correlation can be very useful under certain specific conditions despite the fact that it gives little or no information about the real causes of the sensory quality under consideration. Kramer (31) has commented interestingly on the relevance of correlating objective and subjective data.

Use of computers to accomplish correlation of flavor quality with gas chromatographic data has received considerable attention. Young (32) and Powers (33, 34) successfully employed stepwise discriminant analysis of GC data as an aid in classifying flavor quality of various coffee blends and batches of aged potato chips. Ratios of per cent intensities of GC peaks were more useful than peak heights as "raw GC data" for computer determination of differentiating value.

The most extensive and useful discussion of computer evaluation of gas chromatographic profiles is that by Biggers et al. (35). Methods for unequivocal differentiation between two coffee varieties, independent of their degree of roast, are discussed, and correlations between GC data and sensory evaluations by experts are considered. A coded series of seven coffee blends (each of which contained up to seven different types) which varied widely in flavor quality in the opinions of coffee experts, was differentiated by computer analysis in perfect agreement with the organoleptic ranks of the blends. The authors state that "these methods of differentiation and quality ranking should be readily applicable to other complex natural products, such as essential oils, for instance. In fact, because of its chemical complexity and the dependence of its chemical composition upon the degree of roasting, coffee might have been thought to be one of the materials least amenable to this kind of treatment of analytical data."

Such accomplishments spark the imagination to wonder what remarkable correlations of objective and subjective data may occur in the future!

IV. SOME CORRELATIONS BETWEEN AROMA AND CHEMICAL STRUCTURE

Although sufficient accurate and valid data are not at hand to permit generalized statements or "rules of thumb" for successful correlation of chemical structure and aroma, progress is being made. Because much of this progress occurred in the perfumery field, however, a large portion of the information remains unpublished and is unfortunately unavailable to the scientific community. However, the situation is changing for the better in that more and more information is being published, and for this we are grateful.

Musk aroma is unusual because it is produced by a variety of structural types. There are macrocyclic, nitro-, isochroman, meta-, and ortho-musks (36). Even though many compounds are included within each musk category, Stoll states that perfumers distinguish their aromas easily (37). Theimer and Davies specified structural criteria that must be met for compounds to be categorized as musks (36). Some additional criteria must be found, however, since certain compounds which meet established structural or dimensional requirements exhibit little or no odor. The effect of relatively minor changes in general molecular shape on aroma intensity of various ortho-musks is illustrated in Fig. 8-2 (36).

Wood-amber aroma is strongly exhibited by four of the compounds shown in Fig. 8-3. Stoll showed that the other compounds, only slightly changed in structure, were odorless or weakly odorous. He also compared aromas of the terpenyl-3-cyclohexanols illustrated in Fig. 8-4 (37). As shown, these compounds, which were carefully synthesized by DeMole (38), vary in the degrees to which they exhibit sandalwood aroma. It is not surprising that positional isomers

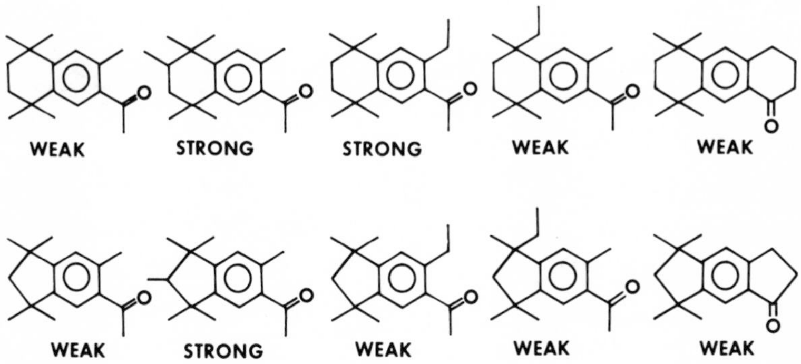

FIG. 8-2   The effect of molecular shape on aroma intensity of certain ortho-musks.

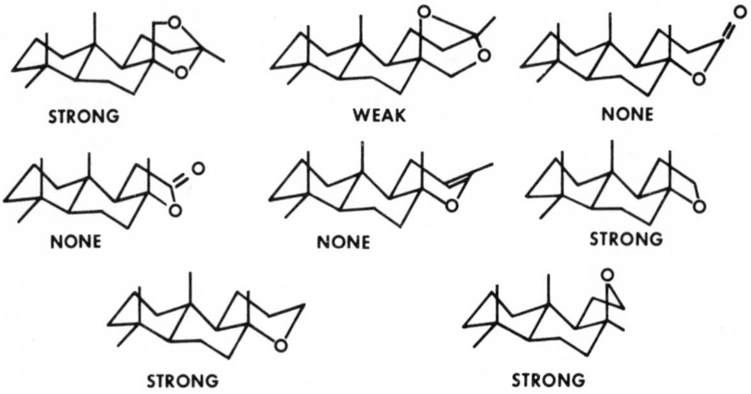

FIG. 8-3   Wood–amber aroma.

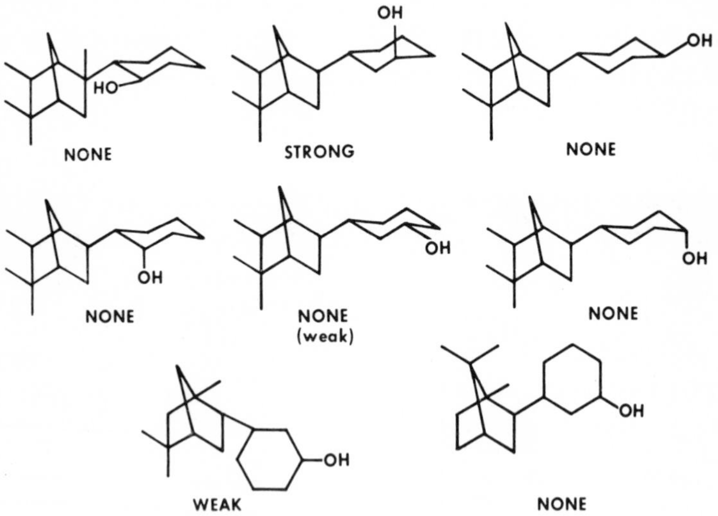

FIG. 8-4   Sandalwood aroma: terpenyl-3-cyclohexanols.

showed different odor qualities. It is striking, however, that when the 3-hydroxyl group is axial, strong sandalwood aroma results, but when the 3-hydroxyl is equatorial, no odor or a very weak odor is produced (37).

Recent elegant syntheses by Pesaro et al. (39) have shown that nootkatone [found in grapefruit and its correct structure assigned by MacLeod (40, 41)] is responsible for the characteristic grapefruit aroma. In the course of the total synthesis of nootkatone, 4-epinootkatone was also synthesized. The epimeric material has no grapefruit aroma (42). The synthesized, racemic nootkatone was indistinguishable from the nootkatone isolated from grapefruit oil with respect to aroma quality and odor threshold (43). Because the synthetic material is identical to the material isolated from grapefruit in chemical, physical, and sensory properties, nootkatone is now known to cause the characteristic odor. The odor is not caused (as feared) by an impurity cocrystallized with nootkatone. The aroma of 4-epinootkatone is not at all similar to that of nootkatone. As was the case with the axial and equatorial 3-hydroxyl groups in the terpenyl-3-cyclohexanols, this is an excellent example of a striking difference in aroma caused by a very small difference in molecular structure.

The smallest known difference between molecules is that between enantiomers. Naves pointed out that in no known case is one enantiomer odorous and the other not odorous (44). Comparisons between enantiomers, and between enantiomers and their racemic form, have always been shadowed by doubt about the sensory purity of the samples. One hopes that modern advances in separation techniques will permit valid comparisons to be made in the near future.

A study by Winter (45) of the homologs and analogs of 1-(p-hydroxyphenyl)-3-butanone ("raspberry ketone"), first isolated by Schinz and Seidel (46) from raspberries, demonstrated that only this one compound exhibited strong raspberry taste and aroma. Its methyl ether and acetate derivatives had weak raspberry flavor, but positional isomers with the hydroxyl group ortho or meta on the ring, or with the carbonyl group in the 1, 2, or 4 position in the butyl side chain, had little or no raspberry flavor. Some homologs having a longer than four-carbon side chain were faintly raspberry while others were not even slightly fruity in character. This systematic study showed that the structure 1-(p-hydroxyphenyl)-3-butanone is highly specific with respect to raspberry flavor and that even small structural changes diminish the flavor in varying degrees.

The characteristic highly potent aroma component of bell peppers, 2-methoxy-3-isobutylpyrazine, was isolated, identified, and synthesized by Buttery et al. (26). Its odor threshold is two parts per 10^{12} parts of water! Months after the synthesis of this component, the laboratory smelled of freshly chopped green bell peppers. Because of this compound's high odor potency and specific characteristic aroma, other 2-methoxy-3-alkylpyrazines were synthesized and their thresholds determined (47). These thresholds, shown in Table 8-13, provide a quantitative measure of the effects of changes in the molecular structure of these compounds on their aroma intensity. Removal of an alkyl side chain, in the case of 2-methoxypyrazine, caused a decrease in odor potency of 3.5×10^5. 2-Isobutyl- and 2,5-dimethylpyrazines are weak odorants in water solution.

Comparison of the quality or type of odor exhibited by the pyrazines listed in Table 8-13 showed that the isobutyl and propyl methoxypyrazines were very much like bell peppers. The isopropyl methoxypyrazine smelled somewhat like raw potato and only moderately like bell pepper. The ethyl methoxypyrazine was like raw potato in aroma and not like bell pepper.

Pyrazines, in general, are of great interest as aroma compounds. They contribute in an important way to the flavor of a variety of roasted, or otherwise cooked, foods, such as coffee (48, 49), cocoa (50), peanuts (51), and potato chips (52).

TABLE 8-13

Odor Thresholds of Certain Pyrazine Derivatives[a]

Compound	Odor threshold in parts of compound per 10^{12} parts of water
2-Methoxy-3-isobutylpyrazine	2
2-Methoxy-3-propylpyrazine	6
2-Methoxy-3-isopropylpyrazine	2
2-Methoxy-3-ethylpyrazine	500
2-Methoxypyrazine	700,000
2-Isobutylpyrazine	400,000
2,5-Dimethylpyrazine	1,800,000

[a]Reprinted from Ref. (47), p. 248.

The presence of methyl propenyl disulfide and of cis- and trans-propenyl disulfides in the green leaves of chives has been suggested by Wahlroos and Virtanen (53). These compounds are also important constituents of onion oil (54, 55). The synthesis of 1-alkenylalkyl disulfides by Mijers et al. (56) verified that these compounds have a powerful odor, reminiscent of the smell of freshly cut onions. These workers also reported that alkylvinyl disulfides have a rather unpleasant odor. It is very well known that diallyl disulfide causes garlic aroma (57) and that diallylthiosulfinate ("allicin") causes "fresh" garlic aroma (58). The characteristic differences in the aromas of these highly potent sulfur compounds are caused by relatively small differences in their molecular structure.

Certain furanones, shown in Fig. 8-5, have very interesting aromas. 4-Hydroxy-2,5-dimethyl-3(2H)-furanone is reported to cause caramel and/or burnt pineapple aroma (59-61); the 4-hydroxy-5-methyl-3(2H)-

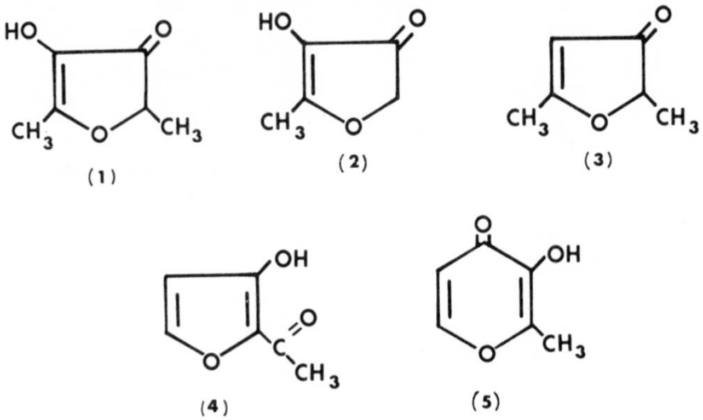

FIG. 8-5   Furanones having aroma importance. (1) 4-Hydroxy-
2,5-dimethyl-3(2H)-furanone. (2) 4-Hydroxy-5-
Methyl-3(2H)-furanone. (3) 2,5-Dimethyl-3(2H)-
furanone. (4) Isomaltol. (5) Maltol.

furanone gives rise to roasted chicory root aroma (62). Both compounds are components believed to contribute to beef broth flavor (62).
2,5-Dimethyl-3(2H)-furanone is said to have the aroma of freshly
baked bread. Isomaltol and maltol, as well as other products of
caramelization and pyrolysis of carbohydrates, are mentioned by
Hodge (63) as having burnt, pungent, fruity, fragrant, and caramel
odors.

For a long time lactones have been known to be important in various
characteristic aromas. Moncrieff (64) summarized the work of
Rothstein (65, 66) on the preparation and description of the α- and
γ-butyrolactones. The γ-lactones with varying alkyl side chains
shown in Fig. 8-6 are of particular interest. As the side chain is
lengthened, the aroma is changed from coconut to peach to peach-musk.
The lactone with R hexyl (C_6H_{13}) was isolated from peaches by
Sevenants and Jennings (67). A lactone, α-hydroxy-β-methyl-
γ-carboxy-Δ^{α}, β-γ-hexenolactone [(7) in Fig. 8-6], isolated from
a protein hydrolysate, exhibits the flavor of beef bouillon (68, 69).

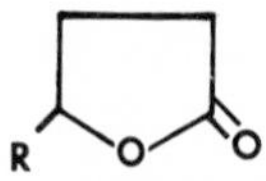

R = C₅H₁₁, COCONUT
C₆H₁₃, PEACH
C₇H₁₅, PEACH
C₈H₁₇, PEACH-MUSK

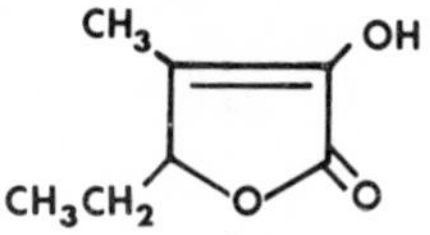

FIG. 8-6   Lactones having aroma importance.

The aroma characteristics of phthalides have been of interest for some time. The celery-like odors of synthesized 3-alkylidene phthalides and hydrophthalides were noted by Berlingozzi (70) and Kariyone and Shimizu (71). Investigations of volatile constituents of celery showed that the following compounds contributed importantly to characteristic celery aroma: 3-isobutylidene-3a-4-dihydrophthalide, 3-isovalidene-3a-4-dihydrophthalide, 3-isobutylidene phthalide, and 3-isovalidene phthalide. The structures of these closely related substances are shown in Fig. 8-7. They exhibit readily detectable celery-like aromas at levels of about 0.1 ppm in water (72, 73). Sedanolide (12) was isolated from celery seed oil many years ago (74).

V. SUMMARY

Additional examples of correlations between molecular structure and sensory quality might be discussed, as could numerous examples in which no correlations have been achieved. When viewed in their broad context, the problems of attaining correlations in flavor research are far from solved. The situation with respect to chemical identification is very good. Use of the powerful new analytical tools in government, industrial, and university research laboratories has generated much interest and much has been accomplished. Pressure is therefore increasing for elucidation of the physiological basis of odor and taste detection and of the psychophysics of flavor evaluation.

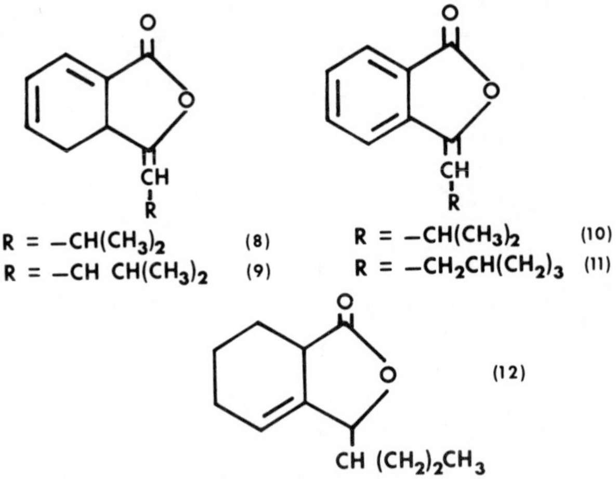

R = −CH(CH₃)₂   (8)
R = −CH CH(CH₃)₂   (9)

R = −CH(CH₃)₂   (10)
R = −CH₂CH(CH₂)₃   (11)

(12)

CH (CH₂)₂CH₃

FIG. 8-7   Some phthalides having celery aroma. (8)
3-isobutylidene-3a-4-dihydrophthalide. (9)
3-isovalidene-3a-4-dihydrophthalide.
(10) 3-isobutylidene phthalide. (11) 3-isovalidene
phthalide.

Major interdisciplinary efforts on the part of physiologists, chemists, psychophysicists, and food scientists are required. These efforts will be fully justified if a basic understanding of flavor as it relates to food acceptance is achieved.

REFERENCES

1. D. G. Moulton, J. Food Sci., 30, 908 (1965).

2. T. H. Schultz, R. A. Flath, D. R. Black, D. G. Guadagni,
    W. G. Schultz, and R. Teranishi, J. Food Sci., 32, 279 (1967).

3. J. A. Swets, Science, 134, 168 (1961).

4. E. Linker, M. E. Moore, and E. Galanter, J. Exptl. Psychol.,
    67, 59 (1964).

5. H. W. Berg, R. M. Pangborn, E. B. Roessler, and A. D.
    Webb, Nature, 157, 108 (1963).

6. H. Stone, J. Appl. Physiol., 18, 746 (1963).

7. D. G. Guadagni, R. G. Buttery, and S. Okano, J. Sci. Food Agr., 14, 761 (1963).

8. S. Patton, J. Food Sci., 29, 679 (1964).

9. W. Stark and D. A. Forss, J. Dairy Res., 31, 253 (1964).

10. D. G. Guadagni, R. G. Buttery, S. Okano, and H. K. Burr, Nature, 200, 1288 (1963).

11. D. G. Guadagni, R. G. Buttery, and J. Harris, J. Sci. Food Agr., 17, 142 (1966).

12. R. G. Buttery, R. M. Seifert, D. G. Guadagni, D. R. Black, and L. C. Ling, J. Agr. Food Chem., 16, 1009 (1968).

13. M. Rothe and B. Thomas, Z. Lebensm. Untersuch. Forsch., 119, 302 (1963).

14. D. G. Guadagni, S. Okano, R. G. Buttery, and H. K. Burr, Food Technol., 20, 518 (1966).

15. R. A. Flath, D. R. Black, D. G. Guadagni, W. H. McFadden, and T. H. Schultz, J. Agr. Food Chem., 15, 29 (1967).

16. D. E. Heinz, R. M. Pangborn, and W. G. Jennings, J. Food Sci., 29, 756 (1964).

17. A. I. McCarthy, J. K. Palmer, C. P. Shaw and E. E. Anderson, J. Food Sci., 28, 379 (1963).

18. E. L. Wick, E. Murray, J. Mizutani, and M. Koshika, Advan. Chem. Ser., 65, 12 (1967).

19. J. L. Bomben, D. G. Guadagni, and J. G. Harris, Food Technol., 23, 83 (1969).

20. J. E. Langler and E. A. Day, J. Dairy Sci., 12, 1291 (1964).

21. P. W. Meijboom, J. Am. Oil Chemists' Soc., 41, 326 (1964).

22. E. S. Keith and J. J. Powers, J. Food Sci., 33, 213 (1968).

23. W. G. Jennings and M. R. Sevenants, J. Food Sci., 29, 158 (1964).

24. W. G. Jennings, R. K. Creveling, and D. E. Heinz, J. Food Sci., 29, 730 (1964).

25.   D. A. Forss, E. A. Dunstone, E. H. Ramshaw, and W. Stark, J. Food Sci., 27, 90 (1962).

26.   R. G. Buttery, R. M. Seifert, R. E. Lundin, D. G. Guadagni, and L. C. Ling, Chem. Ind., 1969, 490.

27.   W. D. MacLeod, Jr., and N. M. Buigues, J. Food Sci., 29, 565 (1964).

28.   C. S. Tang and W. G. Jennings, J. Agr. Food Chem., 15, 24 (1967).

29.   R. Harper, E. C. Bate-Smith, and D. G. Land, Odour Description and Odour Classification, American Elsevier, New York, 1968.

30.   G. A. F. Harrison, The Brewer's Digest, June 1967, p. 74-76.

31.   A. Kramer, Food Technol., 23, 926 (1969).

32.   L. L. Young, M. S. Thesis, Univ. Georgia, Athens, 1968.

33.   J. J. Powers, Food Technol., 22, 39 (1968).

34.   J. J. Powers and E. S. Keith, J. Food Sci., 33, 207 (1968).

35.   R. E. Biggers, J. J. Hilton and M. A. Gianturco, J. Chromatog. Sci., 7, 453 (1969).

36.   E. T. Theimer and J. T. Davies, J. Agr. Food Chem., 15, 6 (1967).

37.   M. Stoll, Revue de Laryngologie, G. Portmann, Bordeaux, 1965, p. 972.

38.   E. Demole, Helv. Chim. Acta, 47, 1766 (1964).

39.   M. Pesaro, G. Bozzato, and P. Schudel, Chem. Commun., 1152 (1968).

40.   W. D. MacLeod, Jr., and N. M. Buigues, J. Food Sci., 29, 565 (1964).

41.   W. D. MacLeod, Jr., Tetrahedron Letters, 4779 (1965).

42.   P. Schudel, Givaudan-Esrolko Ltd., Dubendorf-Zurich, Switzerland, private communication.

43. D. G. Guadagni, Western Regional Laboratories, Albany, Calif.,
     private communication.

44. Y. R. Naves, in Molecular Structure and Organoleptic Quality,
     Society of Chemical Industry, London, 1957, p. 38.

45. M. Winter, Helv. Chim. Acta, 44, 2110 (1961).

46. H. Schinz and C. F. Seidel, Helv. Chim. Acta, 44, 278 (1961).

47. R. M. Seifert, R. G. Buttery, D. G. Guadagni, D. R. Black,
     and J. G. Harris, J. Agr. Food Chem., 18, 246 (1970).

48. H. A. Bondarovich, P. Friedel, V. Krampl, J. A. Renner,
     F. W. Shephard, and M. A. Gianturco, J. Agr. Food Chem.,
     15, 1093 (1967).

49. I. M. Goldman, J. Seibl, I. Flament, F. Gautschi, M. Winter,
     B. Willhalm, and M. Stoll, Helv. Chim. Acta, 50, 694 (1967).

50. J. P. Marion, F. Muggler-Chavan, R. Viani, J. Bricout,
     D. Reymond, and R. H. Egli, Helv. Chim. Acta, 50, 1509
     (1967).

51. M. E. Mason, B. Johnson, and M. C. Hamming, J. Agr. Food
     Chem., 14, 454 (1966).

52. R. E. Deck and S. S. Chang, Chem. Ind., 1965, 1343.

53. O. Wahlroos and A. I. Virtanen, Acta Chem. Scand., 19, 1327
     (1965).

54. A. I. Virtanen, Nutr. Dieta, 9, 1 (1967).

55. Central Institut voor Voedingsonderzoek TNO, Ziest, The
     Netherlands, and Naarden Research Department, Naarden, The
     Netherlands, unpublished results.

56. H. E. Wijers, H. Boelens, A. van der Gen, and L. Brandsma,
     Rec. Trav. Chim., 88, 519 (1969).

57. F. W. Semmler, Arch. Pharm., 230, 434 (1892).

58. C. J. Cavallito and J. H. Bailey, J. Am. Chem. Soc., 66,
     1950 (1944).

59. J. E. Hodge, U. S. Pat. 2,936,308 (1960).

60.   J. E. Hodge, B. E. Fosher, and E. C. Nelson, Am. Soc.
      Brewing Chemists Proc., 1963, p. 84.

61.   R. M. Silverstein, in Symposium on Foods: The Chemistry and
      Physiology of Flavors (H. W. Schultz, ed., E. A. Day and
      L. M. Libbey, associate eds.), Avi, Westport, Conn., 1967,
      p. 450.

62.   C. H. T. Tonsbeek, A. J. Plancken, and T. v. d. Weerdhof,
      J. Agr. Food Chem., 16, 1016 (1968).

63.   J. E. Hodge, in Symposium on Foods: The Chemistry and
      Physiology of Flavors (H. W. Schultz, ed., E. A. Day and
      L. M. Libbey, associate eds.), Avi, Westport, Conn., 1967,
      p. 465.

64.   R. W. Moncrieff, The Chemical Senses, Leonard Hill, London,
      1951, p. 254.

65.   B. Rothstein, Bull. Soc. Chim., 2(5), 80 (1935).

66.   B. Rothstein, Bull. Soc. Chim., 2(5), 1936 (1935).

67.   M. R. Sevenants and W. G. Jennings, J. Food Sci., 31, 81
      (1966).

68.   H. Sulser, Knorr Food Products Company, Zurich, Switzerland,
      private communication, 1969.

69.   H. Sulser, J. DePizzol, and W. Buchi, J. Food Sci., 32, 611
      (1967)

70.   S. Berlingozzi, Gazz. Chim. Ital., 57, 264 (1927).

71.   T. Kariyone and S. Shimizu, J. Pharm. Soc. Japan, 73, 336
      (1953).

72.   H. J. Gold and C. W. Wilson, J. Food Sci., 28, 484 (1963).

73.   H. J. Gold and C. W. Wilson, J. Org. Chem., 28, 985 (1963).

74.   G. Ciamician and P. Silber, Ber., 30, 492, 50k, 1419, 1424,
      1427 (1897).

75.   E. von Sydow, J. Andersson, K. Anjou, G. Karlsson, D. Land,
      and N. Griffiths, Lebensm.-Wiss. U. Technol., 3, 11 (1970).

AUTHOR INDEX

Numbers in brackets are reference numbers and indicate that an author's work is referred to although his name is not cited in the text. Underlined numbers give the page on which the complete reference is listed.

A

Abrahamsson, S., 135[6], 224[52], 177, 231
Adachi, A., 9[19], 32
Albrecht, J.J., 11[26], 32
Amoore, J.E., 16, 33
Amy, J.W., 143, 219[45], 235[13], 178, 231, 255
Anderson, D.F., 19, 20[47], 33
Anderson, E.E., 116[17], 117[17], 274[17], 131, 291
Angelini, P., 48[16,17], 73
Arnold, R.G., 21[50], 114, 33, 131
Arsenault, G., 143[13], 178
Atkinson, E.P., 234[2], 255
Averill, W., 92[21], 105

B

Bailey, G.F., 252[43], 287[58], 257, 293
Baitinger, W.E., 143[15], 235[13], 178, 255
Banner, A.E., 182[1], 228
Barber, M., 183[5], 225[58], 228, 232
Barron, R.P., 151[22], 178
BateSmith, E.C., 277[29], 292

Bazinet, M.L., 27[66], 48[16, 17], 143[11], 169[38], 170[38], 171[38, 39], 219[46], 34, 73, 178, 179, 231
Beckey, H.D., 151[23], 152[23], 178
Beidler, L.M., 3,4,5[11,27], 12, 14, 31, 32
Berg, H.W., 266[5], 290
Berlingozzi, S., 289, 294
Beroza, M., 28[70,71,74], 204[30], 213, 34, 230
Beynon, H.J., 135, 140[7],153, 182[3], 177, 228
Bidmead, D.S., 61[43], 74
Biemann, K., 28[67], 135, 140[3], 143[13], 163[35], 189[13], 190[13], 218[42], 219[47], 220, 224[53], 225[55,57], 234[9], 34, 177, 178, 179, 229, 230, 231, 255
Bierl, B.A., 28[72], 213, 34, 230
Biggers, R.E., 29[77], 100[36], 282, 35, 106, 292
Bills, D.D., 48[18], 73

SUBJECT INDEX

A

Abbreviated mass spectrum, 224
Accelerating electrode, 147, 148
Accelerating voltage, 154-157,
    162, 183
Acetic acid, 10, 20
6-Acetyl-1,1,2,2,3,3,5-hepta-
    methylindane, 14
Activated charcoal, 144
Adam's catalyst, 210
Adsorption, 63, 144, 189, 198
  on charcoal, 63, 67
Adsorption chromatography, on
    silica gel, 63
Aerosol, 234
Aldehydes, from ozonolysis, 213
Alkaloids, 6
Alkanes, recovery from corn oil,
    47, 49
Alkanoates, ethyl, recovery from
    water
  by distillation, 47
  by extraction, 65
Alkanols, recovery from water
  by distillation, 47
  by extraction, 65
2-Alkanones, recovery from
    water
  by distillation, 47
  by extraction, 65
2-Amino-4-nitro-propoxybenzene,
    6
Ammonium chloride, 5
Amplifier, 164, 166, 167-168,
    173, 182
  solid state, 168, 173
n-Amyl acetate, 77
Analog-to-digital converters,
    222

Analyses, special requirements
    of aroma chemistry, 77
Analysis, basic steps, 77, 78
Analyzer, in mass spectrometers,
    139, 153-163
Δ^{16}-Androsten-3-α-ol, 14
Anode, 147
Apple essence, 65, 262, 269
Apple juices, GC, 115, 118
L-arabinose, 11
Argon, 238
Aroma
  bell pepper, 286
  as biological property, 39, 262,
    266
  bread, 288
  celery, 289
  cocoa, 61
  definition, 40
  differences between enantiomers,
    285
  grapefruit, 285
  origin, 37, 38
  retention in food liquids, 57
  retention in "model" food
    matrices, 57
  vapor analysis of, 38, 260-262
Aroma concentrate, definition, 37
Aroma concentrates, 38, 63, 201,
    227, 261, 262, 268
  apple, 65, 67
  banana, 63
  bread pre-ferments, 61
  cocoa, 61
  fruit juices, 61
  molasses, 61
  pear, 61
  peas, 63
  raspberry, 61
Aroma intensity(see Odor intensity)